SpringerBriefs in Materials

W0234541

For further volumes:
http://www.springer.com/series/10111

Nikhil Gupta · Dinesh Pinisetty
Vasanth Chakravarthy Shunmugasamy

Reinforced Polymer Matrix Syntactic Foams

Effect of Nano and Micro-Scale Reinforcement

 Springer

Nikhil Gupta
Mechanical Engineering Department
Polytechnic Institute of NYU
Brooklyn, NY
USA

Dinesh Pinisetty
The California Maritime Academy
Vallejo, CA
USA

Vasanth Chakravarthy Shunmugasamy
Mechanical Engineering Department
Polytechnic Institute of NYU
Brooklyn, NY
USA

ISSN 2192-1091   ISSN 2192-1105   (electronic)
ISBN 978-3-319-01242-1   ISBN 978-3-319-01243-8   (eBook)
DOI 10.1007/978-3-319-01243-8
Springer Cham Heidelberg New York Dordrecht London

Library of Congress Control Number: 2013946314

Printed on acid-free paper

Springer is part of Springer Science+Business Media (www.springer.com)

Preface

Hollow particle-filled composites, called syntactic foams, are also classified as closed-cell foams. Enclosing porosity inside thin stiff shells of particles provides reinforcing effect to every void present in the materials microstructure. Such composites can be tailored to have higher specific modulus than the matrix resin and a high level of energy absorption under compression. Increasing interest of marine and aerospace structures in lightweight composites has generated considerable interest in finding new methods to enhance the properties of syntactic foams, including exploring micro and nanosized reinforcements. Higher gas mileage requirements for automobiles have also pushed them to look for lighter weight materials, where syntactic foams are expected to be useful in some components.

In reinforced syntactic foams, the presence of hollow particles and one or more additional reinforcing phases can generate complex deformation and failure mechanisms. The interaction between the mechanisms contributed by micro- and nano-scale materials can also be complex. Only systematic large-scale experimental studies followed by modeling and simulation efforts can help in decoupling such effects and help in designing effective material microstructures. Often, the incorporation of additional phase may be directed by the desire of obtaining a specific set of properties, which may not be limited to mechanical properties. Carbon nanotube and nanofibers affect mechanical, electrical, and thermal properties of syntactic foams. Such reinforced syntactic foams may be developed in the form of multifunctional materials.

This work summarizes and critically analyzes the progress made in the design and analysis of reinforced syntactic foams. Nanofibers, nanoclay, and microfibers provide different strengthening mechanisms in reinforced syntactic foams. Comparative studies conducted in this book have shown some surprising trends. For example, irrespective of the reinforcement type, the tensile and compressive properties of most reinforced syntactic foams vary linearly with respect to the composite density. It is also shown that the existing theoretical models can be extended to predict the elastic properties of nanoscale reinforced syntactic foams and the results are validated with the experimental data. Discovery of commonality in the experimental trends and applicability of theoretical models can guide future studies and help in understanding the potential for developing syntactic foams for transportation and structural applications.

From a vast body of literature on reinforced syntactic foams, it is possible to miss some contributions in the references. We have primarily covered the information available in journal publications. The field continues to evolve at a rapid pace. Advancements in the understanding of nanomaterials and nanocomposites directly impact the reinforced syntactic foams field. We hope that this brief book will provide a starting point for the interested readers to gain basic understanding about the major material parameters and mechanical properties of reinforced syntactic foams.

Brooklyn, New York, USA   Nikhil Gupta
Vallejo, California, USA   Dinesh Pinisetty
Brooklyn, New York, USA   Vasanth Chakravarthy Shunmugasamy

Acknowledgments

The authors wish to thank a number of people who contributed to the research related to reinforced syntactic foams in their group over the past decade. These individuals include Dr. Nguyen Q. Nguyen, Dr. Dung D. Luong, Dr. Gabriele Tagliavia, Ronald L. Poveda, Tien Chih Lin, Momchil Dimchev, Anton Talalayev, Ryan Caeti, Dennis John, and Gleb Dorogokupets. Diligent work by these current and past students resulted in several publications that became the basis for developing this book. We also wish to thank William Ricci and Dr. Gary Gladysz of Trelleborg Offshore, Boston for information related to applications of syntactic foams and technical discussions.

Parts of this work are supported by the Office of Naval Research grant N00014-10-1-0988, Army Research Laboratory cooperative agreement W911NF-11-2-0096, and the previous National Science Foundation grants. The views expressed in this text are those of authors, not of funding agencies.

Contents

Chapter 1
Introduction

Abstract Lightweight materials are of great interest to transportation applications. Structural weight reduction directly translates into fuel saving and increased payload capacity. Porous materials can provide significant weight saving but their applications are limited by their low strength and modulus. This chapter provides an introduction to porous materials, which includes open- and closed-cell foams. The closed-cell foams can be further divided into foams containing gas porosity and the foams containing hollow particles. The hollow particle filled porous materials are called syntactic foams. These foams are also classified as particulate composites. Enclosure of porosity within a thin but stiff shell helps in obtaining low density in syntactic foams without a severe penalty on the mechanical properties. Syntactic foams possess superior properties under compression compared to foams comprising gas porosity in the matrix. Several micro- and nano-scale reinforcements have been used to improve the tensile and flexural strengths of syntactic foams. Nanoclay, carbon nanofibers, carbon nanotubes, glass fibers, and ceramic particles have been used as reinforcements in syntactic foams. Establishing structure-property correlations of reinforced syntactic foams will pave way to design effective lightweight composites for engineering structures. The chapter also discusses some of the present day applications of syntactic foams.

Keywords Foam • Open-cell foam • Closed-cell foam • Syntactic foam • Porosity • Hollow particle • Marine structures • Thermal insulation • Polymer foam • Porous material

Use of lightweight materials can result in significant saving of energy and resources in a number of fields. For example, weight reduction of automobiles, marine vessels, and aircraft can save fuel, increase payload capacity, and contribute to reducing pollution [1, 2]. Reduction in structural weight is possible by using lightweight porous materials. However, conventionally porosity is viewed as an

N. Gupta et al., *Reinforced Polymer Matrix Syntactic Foams*, SpringerBriefs in Materials, DOI: 10.1007/978-3-319-01243-8_1, © The Author(s) 2013

undesired microstructural feature in materials because it leads to a reduction in mechanical properties and an increase in the possibility of failure. Common examples of porous materials are various kinds of open and closed-cell foams.

1.1 Porosity in Foams

Open-cell foams contain interconnected porosity. An example of open-cell aluminum foam is shown in Fig. 1.1. In these foams, continuous paths of porosity can be found throughout the entire structure. Such foams are useful in heat exchanger applications because they have a very high surface area to volume ratio and a working fluid can be used to very effectively conduct heat from the foam. In contrast to the open-cell foams, the closed-cell foams do not have interconnected porosity. An example of closed-cell polyvinyl chloride foam is shown in Fig. 1.2. In closed-cell foams, the pores are enclosed from all sides. Such foams have higher mechanical properties than the open-cell foams.

Foams are extensively used in packaging applications due to high damage tolerance under compression, as vibration and sound dampers, and in heat exchangers.

Fig. 1.1 An open-cell aluminum foam shown at two different magnifications

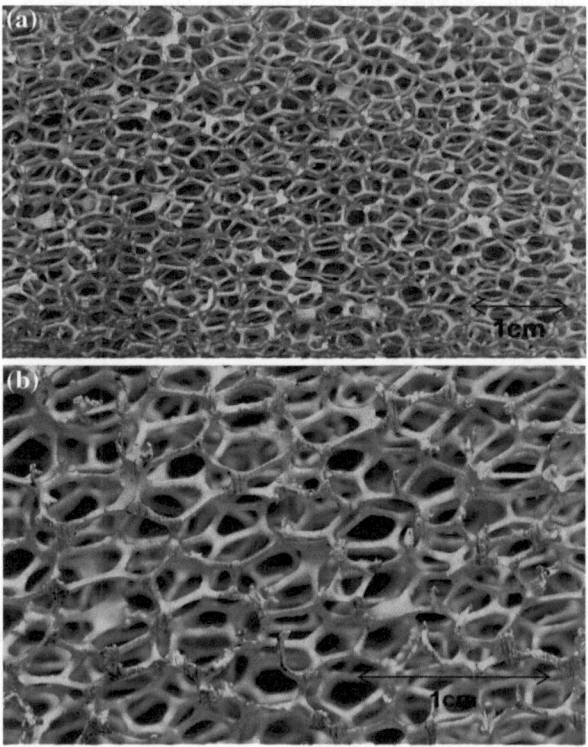

Fig. 1.2 A closed-cell
polyvinyl chloride foam
shown at two different
magnifications

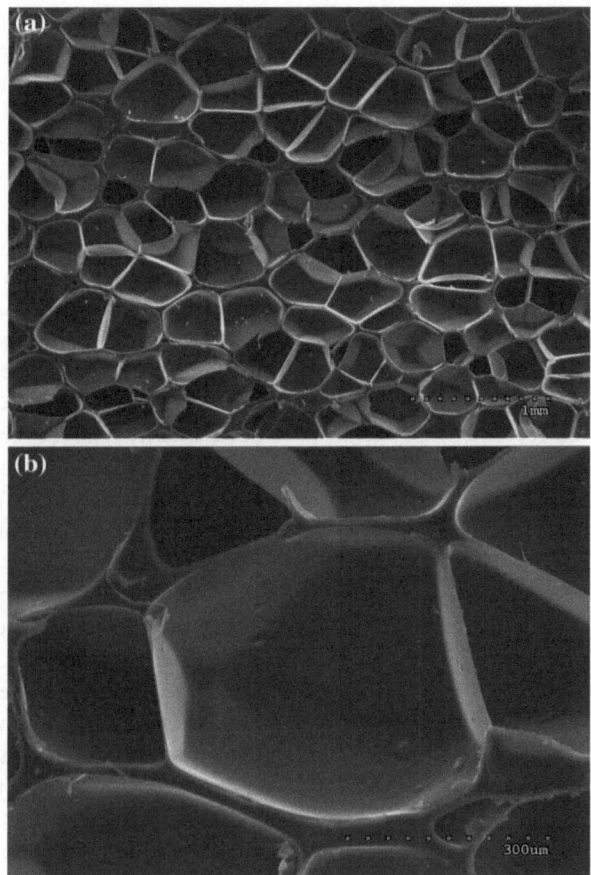

The low mechanical properties of foams are the main limitation in their widespread structural applications. Constructing sandwich structures with foams as core materials can enable their structural applications.

1.2 Syntactic Foams

Innovative methods are required to incorporate porosity in materials to reduce the structural weight without compromising on desired mechanical properties. In this context, a unique class of composite materials, called syntactic foams, is rapidly gaining attention. Syntactic foams comprise hollow particles (also widely referred to as microballoons) filled in a matrix material, resulting in a lightweight composite microstructure. The porosity enclosed inside thin stiff shells of hollow particles provides them with a closed-cell foam-like structure [3, 4]. The mechanical properties such as modulus and strength of syntactic foams are found to be better than the foams comprising gas porosity in the matrix material [5]. The lightweight combined with low moisture absorption makes syntactic foams suitable for fabricating floatation devices [6] and submarine buoyancy systems [7].

1.3 Reinforced Syntactic Foams

Addition of micro and nano-scale reinforcements can help in obtaining properties that are beyond those obtained in plain syntactic foams, which contain only hollow particles in a matrix resin. Epoxy, vinyl ester, and phenolic resins are widely used for fabricating reinforced syntactic foams. Recent advances in the science and technology of nanomaterials have especially benefitted the field of multiscale reinforced composites. This brief critically analyzes the structure and properties of nano and micro-scale reinforced polymer matrix syntactic foams. Nanoscale materials are revolutionizing several engineering fields. Their use in fabricating composite materials has resulted in novel properties and phenomena which may not be feasible in microscale composites. The increasing use of nanoscale reinforcements in syntactic foams requires examining the structure–property relations for various types of reinforced syntactic foams and identifying trends that can help in designing lightweight composites for engineering structures.

This book identifies the commonality in the mechanical property trends in micro and nano-scale reinforced syntactic foams at various syntactic foam compositions. Selection of type and volume fraction of matrix polymer and reinforcement can be guided by the comparative studies. The text also highlights potential directions for future work related to reinforced syntactic foams based on the analysis of the published literature and potential applications of such lightweight composite materials.

1.4 Applications of Syntactic Foams

Syntactic foams are now used in a wide variety of applications. Most of the applications are related to the marine environment, where structural designs depend on the buoyancy obtained from lightweight materials with high compressive strength and modulus. Over the years, the thermal insulation properties of syntactic foams have also been tailored and utilized in industrial applications such as insulation for pipelines, along with other applications in the oil and gas industry. Examples of current applications are discussed in this section.

Buoyancy module for marine applications is the most common application of syntactic foams. Almost all companies that manufacture syntactic foams have reported use of their products in underwater vehicles and ships. Trelleborg Emerson and Cuming (TEC) reports use of syntactic foam blocks in the forward and aft free-flood areas of submarines. Remotely or human-operated vehicles

(ROV and HOV) used in deep sea exploration have been constructed using syntactic foams. The Alvin HOV used for Titanic exploration utilized syntactic foam buoyancy aids. The Alvin HOV, shown in Fig. 1.3, and the Jason ROV, shown in Fig. 1.4, are used by the National Oceanic and Atmospheric Administration (NOAA) for underwater explorations. TEC also reports the use of their syntactic foams in eyebrows for 688 class U.S. nuclear submarines due to their buoyancy, acoustic profile, and ability to significantly improve sonar functions. The control surfaces of this submarine were also filled with a variety of ultralight syntactic foams. The Deepsea Challenger, used by James Cameron for exploration of Mariana Trench, was made of reinforced syntactic foam containing glass fibers and hollow particles in an epoxy resin matrix. This syntactic foam was specifically designed for this submarine that went to the deepest part of the ocean. Utility Development Corporation (UDC) reported use of their syntactic foams in ROVs and autonomous underwater vehicles (AUVs), deep water moorings, torpedo

Fig. 1.3 HOV Alvin used for deep sea exploration. Photo courtesy NOAA [8]

Fig. 1.4 ROV Jason used for deep sea exploration. Photo courtesy NOAA [8]

target, and sonar arrays. UDC also reported the use of their syntactic foams by NOAA in acoustic Doppler current profilers used in marine environments. Engineered Syntactic Systems (ESS) have reported use of their syntactic foam fillers in ship and submarines at Point Loma, Norfolk Naval Shipyard, Northrop Grumman at Newport News, Puget Sound Naval Shipyard, Kings Bay Shipyard, and at the Subsea Naval Base in Groton. ESS syntactic foams were also used in Hydroid REMUS 6000 AUVs used in the Titanic, Amelia Earhart, and Air France 447 search and recovery expeditions. ESS reported use of over 3,500 cubic feet of their syntactic foams in the critical joint areas of the deckhouse of USS Zumwalt (DDG 1000) Guided Missile Destroyer. The image of the deckhouse is shown in Fig. 1.5. ESS supplied syntactic foams have been used on saturation diving bells for thermal and acoustic insulation. Added effect of buoyancy is another useful feature of these syntactic foams. Flotation modules manufactured by Flotation Technologies are shown in Figs. 1.6 and 1.7.

Trelleborg Offshore, Boston (TOB) has an Ecolite product line, which consists of thermoplastic matrix syntactic foams used as a plug assist. High-dimensional stability at high temperatures, stability for a large number of thermal cycles, and easy machining are useful characteristics for this application. A grade of Ecolite is U.S. Food and Drug Administration (FDA) approved for use in food and beverage containers.

Among other applications, UDC reported use of polyurethane-based syntactic foams as insulations for pipelines. The company also makes a class of syntactic foams having low dielectric properties for microwave insulation applications. CRG Industries have suggested use of their syntactic foams for small arms grips and rifle stocks, apart from the marine and aerospace structural applications. Precision Acoustics supplies syntactic foams for acoustic transducers. One of the applications of syntactic foams that gained worldwide fame was its utilization in soccer balls that were used in the 2006 World Cup. The technology developed for this ball by Bayer, relied on a polyurethane syntactic foam. The syntactic foam layer helped in regaining the spherical shape of the ball immediately after being kicked, thereby enabling it to travel in the precise intended trajectory.

Fig. 1.5 Deckhouse of USS Zumwalt at Norfolk Naval Station [9]

Fig. 1.6 Flotation modules
used in marine applications
a a batch of modules
and **b** deployment of a
module. Images courtesy:
Peter Russell, Flotation
Technologies

Fig. 1.7 Flotation riser
modules used for marine
pipelines. Image courtesy:
Peter Russell, Flotation
Technologies

Low coefficients of thermal expansion, coupled with low weight, makes syntactic
foams useful as tooling material in composites industry. Ease of machining makes
it fast and inexpensive to machine syntactic foam tooling components that can be
used for laminate fabrication. Thermal insulation properties of syntactic foams

were utilized in space shuttle applications by the US. Syntactic foams were used as the insulation for the external fuel tank and solid rocket boosters. The Air Force Research Laboratory reported the use of carbon nanofibers and other nanomaterial reinforced syntactic foams for space mirrors. Zero or negative thermal expansion coefficient can be achieved in these materials, which is useful in avoiding geometrical distortion. In addition, carbon nanofibers were used to obtain better thermal conductivity in syntactic foams, which helped in maintaining uniform temperature.

The examples show that a diverse set of applications exists for syntactic foams ranging from deep sea vehicles to space shuttles. The applications are expected to grow rapidly as the science and technology related to syntactic foams continues to develop. Theoretical models capable of predicting the modulus of syntactic foams accurately are now available. These models can help in identifying the parameters that can provide syntactic foams of desired properties. Development in the field of nanomaterials, reduction in the cost of carbon nanofibers and nanotubes, development of methods of dispersing nanomaterials in polymer resins, and availability of a wide variety of hollow particles are proliferating the industrial use of syntactic foams.

Chapter 2
Fillers and Reinforcements

Abstract Syntactic foams are two component materials consisting of matrix resin and hollow particles. Reinforced syntactic foams contain an additional reinforcing material. The density of syntactic foams can be tailored based on the appropriate selection of hollow particle density and volume fraction. Glass hollow particles have been a widely used filler material because their low thermal expansion coefficient provides syntactic foams with high dimensional stability. Hollow fly ash particles, called cenospheres, have also been used as filler material. Fly ash cenospheres are inexpensive and help in developing low cost syntactic foams. However, usually defects are present in the walls of cenospheres and the mechanical properties of cenosphere-filled syntactic foams are not as high as those filled with glass hollow particles at the same density level. Enhancement of mechanical properties of syntactic foams beyond those obtained by tailoring the matrix and hollow particles can be obtained by micro- and nano-scale reinforcement of the matrix material. This chapter discusses hollow particle parameters such as wall thickness and density. Structure and properties of various reinforcements including glass fibers, nanoclay, carbon nanofibers (CNFs), carbon nanotubes (CNTs), and rubber particles are also discussed.

Keywords Syntactic foam • Porosity • Hollow particle • Silicon carbide hollow particle • Alumina hollow particle • Microballoon • Microsphere • Carbon nanofiber • Carbon nanotube • Crumb rubber • Nanoclay • Reinforced foam • Glass fiber • Carbon fiber

2.1 Type of Particles

Information about several types of ceramic hollow particles can be found in a review article [10]. Engineered hollow particles of glass, carbon, and phenolic resin have been used in fabricating syntactic foams [4, 11, 12]. The most widely used hollow

N. Gupta et al., *Reinforced Polymer Matrix Syntactic Foams*, SpringerBriefs in Materials, DOI: 10.1007/978-3-319-01243-8_2, © The Author(s) 2013

Fig. 2.1 Engineered spherical hollow particles of glass

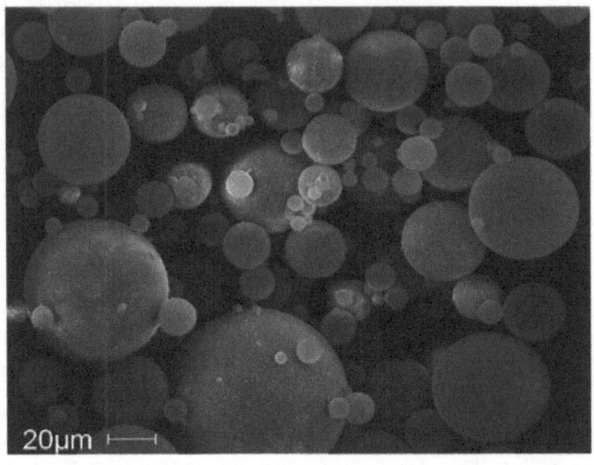

Fig. 2.2 Engineered hollow particles: **a** *cubic-shaped* particles of alumina and **b** *cuboid-shaped* particles of silicon carbide. Particles for imaging are provided by Mr. Oliver M. Strbik III of Deep Springs Technologies, OH

particles have spherical shape with diameter and density in the ranges 10–250 μm and 150–500 kg/m^3, respectively. These low density particles lead to lightweight composites because polymers used as the matrix have densities around 1,100 kg/m^3. A representative sample of 3 M Scotchlite glass hollow particles is shown in Fig. 2.1. These particles are spherical in shape and the particle size can vary over almost two orders of magnitude as observed in this figure. Now engineered hollow particles of almost any shape such as cuboidal or cylinder are available. However, such particles have not been found used in existing syntactic foams. The examples of cubic and cuboid shaped particles of alumina and silicon carbide, respectively, are shown in Fig. 2.2. Specialized applications can use such particles in syntactic foams.

Hollow fly ash particles, called cenospheres, are also used in several studies [2, 13, 14]. Cenospheres are obtained as a byproduct of coal firing in thermal power plants and have a predominantly alumino-silicate composition, along with a large number of impurities and trace elements [15, 16]. Cenospheres are recovered from waste byproducts and show a wide variation in their structure and properties depending on the composition of coal and the firing conditions. Defects such as embedded porosity in the walls can be seen in some fly ash particles in Fig. 2.3. Such defects adversely affect their mechanical properties and make them prone to failure [17, 18]. Compared to cenospheres, engineered hollow particles have better mechanical properties due to their controlled and better material quality. In addition, engineered hollow particles are manufactured under controlled conditions and show better strength and fewer defects in their structure than cenospheres. A possible method to improve the structure and properties of cenospheres is to use them as templates and coat them with a controlled outer layer [19]. The coating can cover defects, make the particle surface more uniform, and improve their mechanical properties. Both cenospheres and engineered hollow particles are widely used in metal and polymer matrix syntactic foams [20–23]. This book mainly covers engineered hollow particle filled polymer matrix syntactic foams because most studies using cenospheres have not reported properties such as

Fig. 2.3 Fly ash cenospheres

density and wall thickness of particles. In the absence of these parameters it is difficult to develop structure-property correlations and apply theoretical models.

The density of syntactic foams can be tailored by selecting the appropriate hollow particle density and volume fraction. In order to obtain properties of syntactic foams beyond those possible by tailoring the matrix and hollow particles, micro and nanosized reinforcements have been used. Carbon nanofibers (CNFs), carbon nanotubes (CNTs), and nanoclay are now widely used to enhance the mechanical properties of polymers [24–29]. Use of such reinforced polymers as matrix can provide syntactic foams with improved mechanical properties. The advantages, processing difficulties, and challenges in attaining the desired properties in reinforced syntactic foams are discussed in detail in the following sections.

2.2 Hollow Particle Parameters

The size and material density are the two main parameters in spherical solid particles. However, analysis of hollow particles requires additional parameters. Figure 2.3 shows that the diameter of hollow particles within a sample can vary over a large range, but the outer diameter does not provide any indication of the particle wall thickness. In the schematic shown in Fig. 2.4a, it can be noted that the particles of the same outer diameter can have different wall thicknesses (t). The particle wall thickness can be defined using a parameter named radius ratio [30]

$$\eta = \frac{R_i}{R_o} \qquad (2.1)$$

(a)

(b)

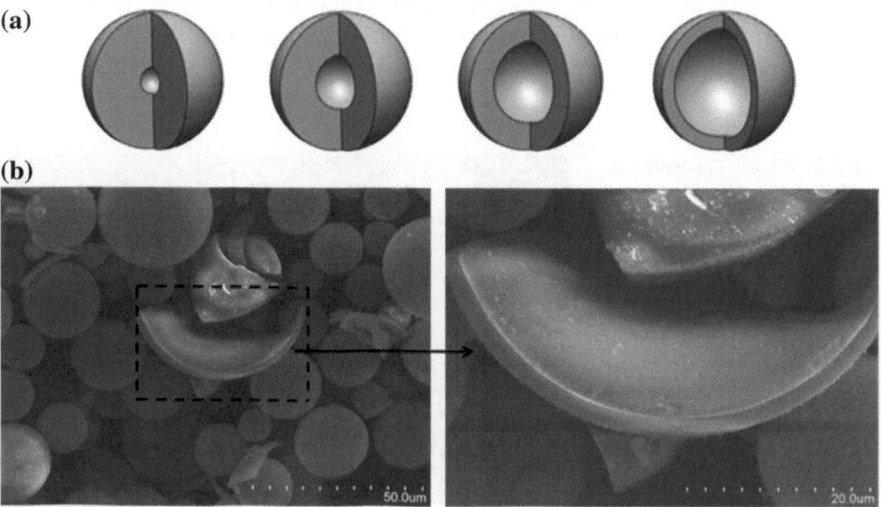

Fig. 2.4 a Schematic of possible variation in wall thickness of hollow particles of the same outer radius and **b** a broken hollow particle showing wall thickness

where R_i and R_o are the internal and outer radii of hollow particles, respectively. Radius ratio can be related to wall thickness by $t = R_o (1-\eta)$.

A combination of radius and wall thickness determines the properties of hollow particles. Direct measurement of wall thickness or radius ratio is usually not practical because they can be different for every particle. Average values of η and t for a batch of particles can be estimated through measurement of true particle density of hollow particles (ρ_{mb}) and the density of the particle material (ρ_g) by

$$\eta = \left(1 - \frac{\rho_{mb}}{\rho_g}\right)^{1/3} \qquad (2.2)$$

where ρ_{mb} and ρ_g can be experimentally measured using a pycnometer. In general, thin-walled particles are used in fabricating syntactic foams. The commonly used glass hollow particles have true particle density in the range 150–500 kg/m^3. Considering the glass density as 2,540 kg/m^3, the radius ratios of hollow particles of densities 150 and 500 kg/m^3 are calculated as 0.93 and 0.98, respectively, which is in a narrow range. The average wall thickness for these particles can be calculated as 0.5 and 1.8 μm, respectively, for commonly used particles of 25 μm radius. Figure 2.4b shows a broken particle of about 25 μm radius having wall thickness of about 2 μm. The wall thickness to diameter ratio is less than 5 % in these particles. Although the radius ratio and wall thickness seem to be in a narrow range, this distribution is wide enough to show different trends in their mechanical properties. Within this radius ratio range, the low density hollow particles cause weakening and the high density particles cause strengthening of syntactic foams as the hollow particle volume fraction is increased [31]. Use of particles with lower radius ratio than 0.93 is not common because no additional benefit in mechanical properties with respect to the density is observed in syntactic foams. In addition, high density foams would not be useful in marine applications that require buoyancy.

It should also be noted that the wall thickness may not be constant within one particle. Locations where walls are thin become weak spots in the composite where failure can initiate. Compression testing of individual hollow particles using a nanoindenter has shown a wide range of mechanical properties in particles of the same size, which may be due to their different wall thicknesses and possible presence of defects in the walls [18].

2.3 Reinforcements

Hollow particles are an essential constituent in the syntactic foam microstructure. Therefore, any second phase material apart from hollow particles incorporated in the matrix is termed as reinforcement in this book. The purpose of incorporating reinforcement includes increase in modulus, strength, and energy absorption; modulation of electrical properties; and tailoring of thermal properties such as thermal expansion, thermal conductivity, and glass transition temperature (T_g). The following discussion presents an overview of different types of reinforcements used in syntactic foams.

2.3.1 Fibers

High aspect ratio glass, carbon and aramid fibers are used in syntactic foams as reinforcements. These fibers have diameter in the range of 8–15 μm, which is of the order of hollow particle diameter. Since the length of fibers can vary over a large range (from less than 1 mm to any desired length), they are termed as microscale reinforcements (or microfibers) based on their diameter measurement. Fibers can be used either in continuous or discontinuous form.

2.3.2 Nanoclay

The schematic structure of nanoclay and high-resolution TEM image of 2 wt% nanoclay dispersed in epoxy resin are shown in Figs. 2.5a and b, respectively [32]. Nanoclay has a stacked platelet structure in which platelets are bonded with each other by van der Waals forces. Intercalation and exfoliation of nanoclay clusters can expose their large surface area to bond with the matrix resin and provide improvement in mechanical properties. Functionalization of nanoclay can make it compatible with different polymers and help in obtaining exfoliation [33]. Shear mixing and ultrasonication are found effective in exfoliating nanoclay in polymeric resins [34]. A vast body of literature is now available on the processing methods for effectively dispersing nanoclay in polymers and properties of nanoclay reinforced composites [35–41].

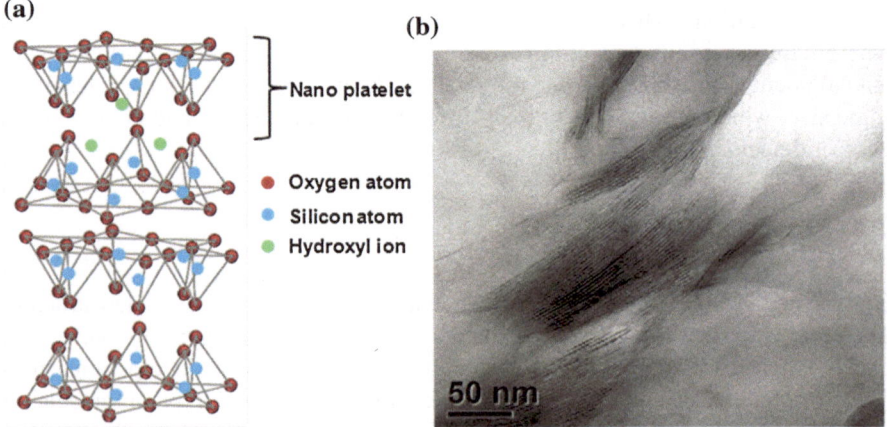

Fig. 2.5 **a** Schematic representation of nanoclay and **b** TEM image of 2 wt% nanoclay dispersed in epoxy resin [32]

2.3.3 Carbon Nanotubes and Nanofibers

CNTs and CNFs have been used as reinforcements in polymer matrix syntactic foams [42]. CNTs and CNFs are long aspect ratio fibrous materials; both contain hollow core structures, and have low density. Entanglement of such long aspect ratio reinforcements is an issue during processing. Use of excessive shear forces to disperse them can lead to their breakage. A representative structure of a CNT (single walled) and a high-resolution TEM image of CNTs suspended in a solution are shown in Figs. 2.6a and b, respectively [43]. Single and multi-wall CNTs are now commercially available. CNTs can be synthesized in desired length and chirality with the available synthesis methods. Several review articles are available about the structure, properties, and applications of CNTs [44–50]. CNT reinforced polymer matrix composites have also been widely studied in recent years and detailed information on these nanocomposites can be obtained from published review articles [51–59].

CNFs are also called stacked-cup CNTs, which refer to the stacked truncated cone structure giving them appearance of a helically coiled graphene ribbon. This kind of structure results in a hollow fiber core and graphene layers oriented at an angle from the fiber axis. The schematic of stacked-cup structure and a TEM image of a CNF sample are shown in Figs. 2.7a and b, respectively. The hollow structure of CNFs can be noticed in Fig. 2.7b. The bonding between graphene layers is important in determining the level of mechanical property improvement obtained in CNF reinforced composites. CNFs have been used for tailoring the mechanical, electrical, and thermal properties of syntactic foams [60]. Detailed discussion on the structure and properties of CNFs [48, 61] and CNF reinforced composites [62–69] can be found in recent review articles.

Theoretical estimates and experimental measurements of mechanical properties of CNTs and CNFs are available in literature. The modulus of CNTs is measured to be over 4 TPa in some studies and is routinely reported in the range 0.95–1.28 TPa [70]. However, the measured elastic modulus of CNT reinforced composites is lower than

(a)   **(b)**

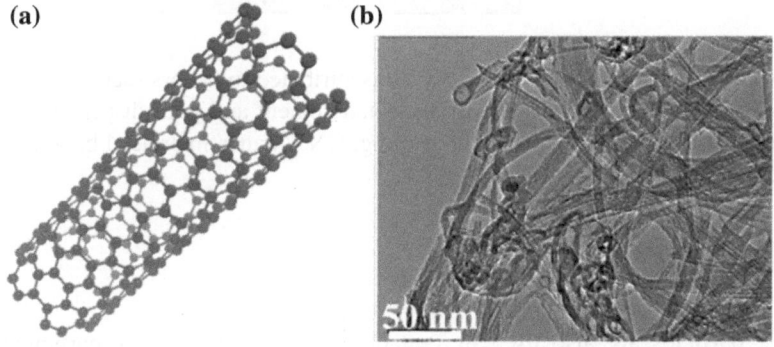

Fig. 2.6 **a** Schematic representation of an armchair carbon nanotube and **b** TEM image of single walled CNTs dispersed in a solution [43]

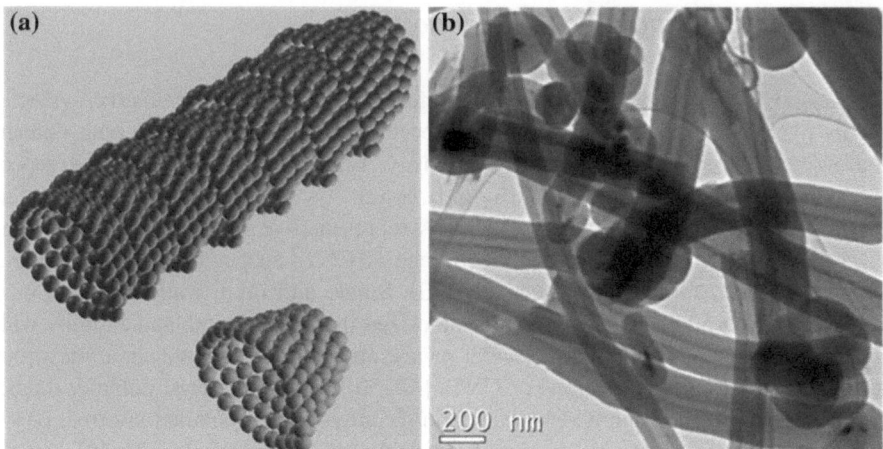

Fig. 2.7 a Schematic representation of stacked-cup structure and **b** TEM image of carbon nanofibers showing hollow core structure

Fig. 2.8 Scanning electron micrograph of crumb rubber particles generated by mechanical grinding process

that expected from such high values, which is attributed to factors such as curviness of CNTs leading to buckling and shearing, entanglement and poor dispersion, porosity entrapped in the composite during processing, CNT/matrix interfacial bonding issues, and defects in CNT structures.

2.3.4 Rubber Particles

Rubber particles are also used in syntactic foams as secondary particulate fillers. Crumb rubber particles obtained from waste tires are used for this purpose [71–74]. Some of these efforts are oriented toward developing useful applications of about 300 million waste tires generated each year in the United States. About 25 % scrap tires are disposed of in landfills, which are an enormous economic and environmental burden [75, 76].

Two methods are used for grinding waste tires to generate crumb rubber particles [77]. In the cryogenic method, the tires are cooled below their embrittlement temperature and then fractured either by impact or grinding. In the second method, the tires are mechanically ground at room temperature. The surface area of room temperature processed particles is higher and they show better reinforcing capabilities in composites. A sample of crumb rubber particles obtained by the mechanical grinding process is shown in Fig. 2.8. A very high surface area and irregular shape of this particle due to shearing fracture are useful in obtaining interfacial bonding and mechanical interlocking with the matrix resin.

Chapter 3
Processing and Microstructure of Syntactic Foams

Abstract This chapter discusses processing methods for reinforced syntactic foams and the effect of processing parameters on the structure and properties of syntactic foams. Enhancement of the mechanical properties of syntactic foams can be achieved by incorporation of micro- or nano-scale reinforcements into the matrix material. Dispersion of nanoparticles and nanotubes and nanofibers in polymer resins is challenging. Mechanical, shear, and ultrasonic mixing techniques have been used for obtaining wetting and dispersion of nanoscale reinforcement in the matrix. Nanoparticle reinforcement may also provide unintentional effect of increased matrix porosity by stabilizing gas bubbles in polymer matrix, if the processing method is not carefully designed. The processing methods are also required to be efficient in promoting wetting of reinforcement by the matrix resin, breaking clusters without fracturing the reinforcement material, and obtaining uniform distribution of reinforcement in the matrix resin. In addition, the hollow particles should not be excessively fractured during the manufacturing process. This chapter provides an overview of various processing methods and the issues encountered during fabrication of reinforced syntactic foams.

Keywords Syntactic foam • Hollow particle • Reinforcement • Nanoscale material • Carbon nanofiber • Carbon nanotube • Porosity • Matrix void • Stir mixing • Agglomeration • Exfoliation • Dispersion

3.1 Processing Methods and Challenges

A commonly used fabrication method for reinforced syntactic foams is illustrated in Fig. 3.1. In this method, a three-step mixing process is used. In the first step, the reinforcement is added to the neat resin and mixed. After thorough mixing of reinforcement, hollow particles are added and stirred until a slurry of consistent viscosity is obtained. In the final step, the hardener or catalyst is added to the resin

N. Gupta et al., *Reinforced Polymer Matrix Syntactic Foams*, SpringerBriefs in Materials, DOI: 10.1007/978-3-319-01243-8_3, © The Author(s) 2013

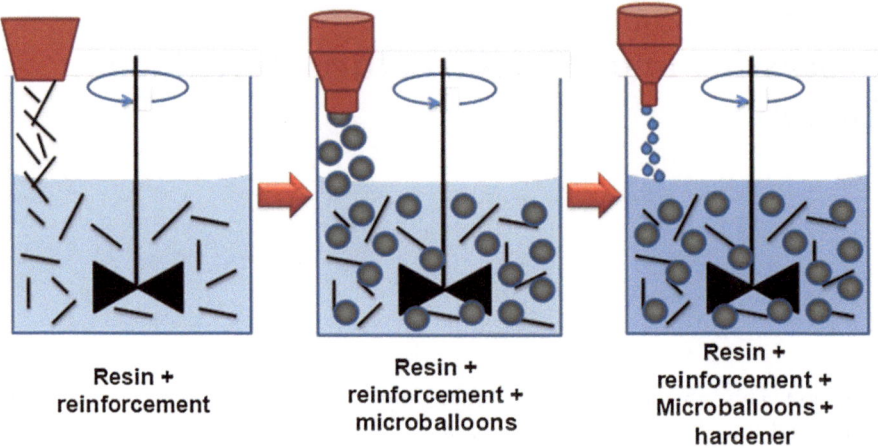

Fig. 3.1 Illustration of reinforced syntactic foam fabrication method

and stirred slowly. The mixture is cast in molds and cured as per the requirements of the resin. Additional rigorous mixing of reinforcement before hollow particles helps in reducing the possibility of hollow particle breakage during processing.

Microfibers can be dispersed relatively easily in polymeric resins using stir mixing methods, but nanoscale fillers require a more rigorous mixing routine. Mechanical mixing using high shear impellers, shear mixing using three-roll mills, and ultrasonic mixing have been widely used to disperse nanoscale materials in polymeric resins [34, 78]. Intercalation and exfoliation of nanoclay in polymeric resins requires high shear mixing techniques [34, 38], which can cause breakage of hollow particles if they are already present in the resin. In addition, incorporation of nanoscale fillers increases the viscosity of the resin [79], which can make it difficult to mix high volume fraction of hollow particles. Long aspect ratio nanostructures are preferred for enhancement in mechanical properties of composites but the dispersion of such materials becomes difficult due to entanglement [80].

A limitation of stirring methods is the volume fraction of hollow particles that can be incorporated in syntactic foams. Since hollow particles have 2–8 times lower density than the matrix resin, they tend to float and segregate in the top part of the foam slab during curing, if less than 30 vol. % particles are incorporated. On the contrary, mixing over 60 vol. % hollow particles is difficult because particles tend to break during stirring and their wetting and clustering become issues during processing. Therefore, syntactic foams with particle volume fraction in the range 0.3–0.6 have been fabricated in most studies.

Reduction in viscosity of the resin can help in mixing high volume fraction of hollow particles without damaging them. Commonly, the resin is heated above room temperature to reduce the viscosity. Another viable option is to add a suitable diluent to the resin. The choice of diluent should be made carefully because many diluents cause a reduction in the modulus of the resin. In some previous studies using epoxy resins as the matrix material, C_{12}-C_{14} aliphatic glycidyl ether

was used as the diluent [81]. Fabrication of reinforced syntactic foams can especially benefit from the choice of diluents and the possibility of reducing the resin
viscosity. A combination of high temperature and diluent is often used to reduce
the viscosity of the resin.

Apart from stir mixing, buoyancy based methods have been used for fabricating
plain syntactic foams [13, 82], but they cannot be used for microfiber reinforced
syntactic foams because the presence of fibers hinders the natural buoyant behavior of hollow particles and may produce localized entrapment and segregation of
hollow particles depending on the fiber volume fraction.

Pressure or vacuum infiltration methods such as resin transfer molding or
vacuum assisted resin transfer molding can be developed for reinforced syntactic
foams, where either a particle preform or a particle/fiber preform is infiltrated by
the resin. However, such methods have not yet been explored for reinforced syntactic foams. Development of particle preforms with higher than 60 vol. % hollow particles can help in fabricating lighter syntactic foam compositions than what
is possible by stirring methods. Infiltration methods can also lead to much higher
loading of fibers than that possible through the mixing methods. In addition, direc-
tionality of fibrous reinforcement can be maintained while infiltrating a preform
and new varieties of reinforced syntactic foams can be developed.

3.2 Microstructure and Porosity Issues

Plain syntactic foams are two component systems, composed of hollow particles
and matrix resin. The microstructure of syntactic foam schematically represented
in Fig. 3.2a, where the porosity is enclosed inside hollow particles. This form of
porosity, termed as "hollow particle porosity," is desired and can be controlled by
means of particle wall thickness and volume fraction [83]. During the fabrication
of syntactic foams, air voids are entrapped in the matrix resin and are termed as
"matrix porosity." The presence of matrix porosity provides a three-phase microstructure to syntactic foams [84]. The matrix porosity is usually undesired and
needs to be minimized through improved process control. Most published studies
have shown up to 5 vol. % matrix porosity (occasionally up to 10 vol. %) in syntactic foams [30, 85–87]. The microstructure of plain syntactic foam containing
only glass hollow particles in the vinyl ester matrix is shown in Fig. 3.2b. Some
matrix voids are also visible in this figure. Use of a shaker or application of vacuum during curing can reduce the matrix porosity. This level of matrix porosity is
not a major concern for current applications that use these materials under compression, and compaction of matrix pores under compression leads to densification
of the foam microstructure. Matrix pores are in the form of a closed-cell structure
and do not lead to any significant moisture uptake if they are uniformly distrib-
uted in the microstructure and are not interconnected. In several instances, matrix
porosity is used as a means to reduce the density of syntactic foams by creating
a three-phase microstructure. However, presence of matrix porosity is a concern

Fig. 3.2 a Schematic
representation of phases
present in syntactic foam
microstructure and
b scanning electron
micrograph of glass hollow
particle/vinyl ester syntactic
foam

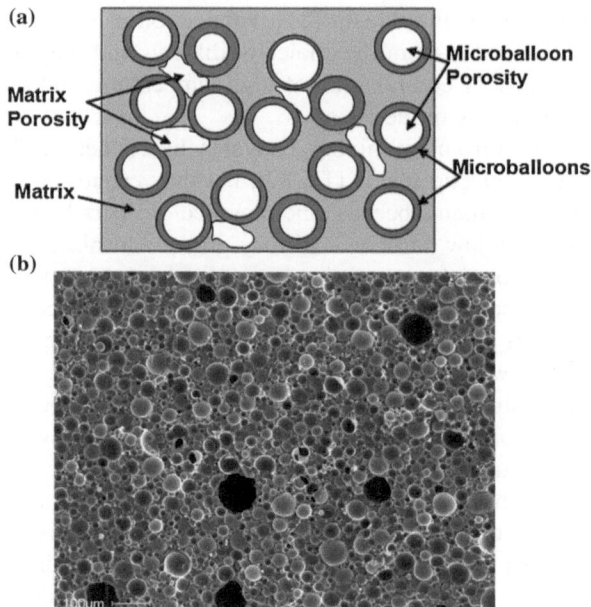

Fig. 3.3 Scanning electron
micrograph of a glass hollow
particle/epoxy syntactic foam
reinforced with randomly
dispersed short glass fibers

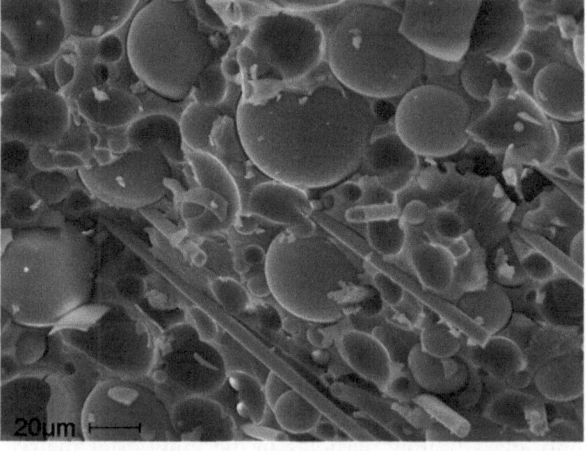

for mechanical properties and high temperature dimensional stability of syntactic
foams and control over the porosity level is desired.

The microstructure of glass fiber reinforced syntactic foam is shown in Fig. 3.3,
where short fibers are randomly dispersed in epoxy resin matrix. The effect of
processing methods on matrix porosity of fiber reinforced syntactic foams has
been studied. The void entrapment in the matrix depends on the mixing proce-
dure employed during manufacturing process and the wetting of hollow particles
and fibers by the matrix medium. Ultrasonic imaging, a non-destructive testing

technique, has been used on fiber reinforced syntactic foams to correlate the manufacturing process with the matrix porosity [88]. Complete mixing of fibers with the matrix material before the addition of hollow particles ensures a more homogeneous microstructure of the foam. In syntactic foams containing 6 mm long chopped glass fibers, 4–11 vol. % matrix porosity was reported [89], which is similar to the matrix porosity levels reported in plain syntactic foams.

Nanoscale reinforcements such as nanoclay or CNF are increasingly used in recent years. Since the commonly used hollow particles are in the size range of

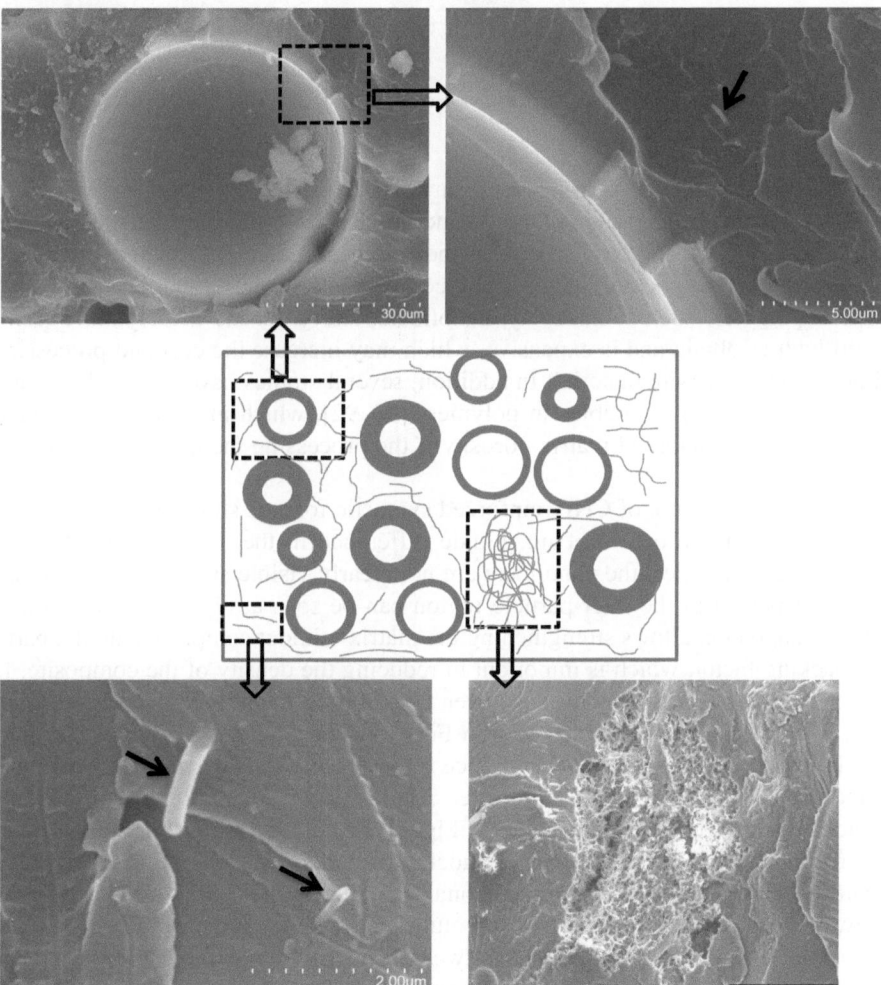

Fig. 3.4 Scanning electron micrographs of various areas in a CNF reinforced syntactic foam. CNFs are marked by solid black arrows in micrographs. A large CNF cluster is shown in the bottom right micrograph, where CNFs have not dispersed well in the matrix resin. A specimen that fractured under tensile loading is used for imaging

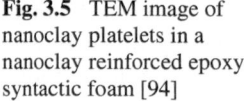

Fig. 3.5 TEM image of nanoclay platelets in a nanoclay reinforced epoxy syntactic foam [94]

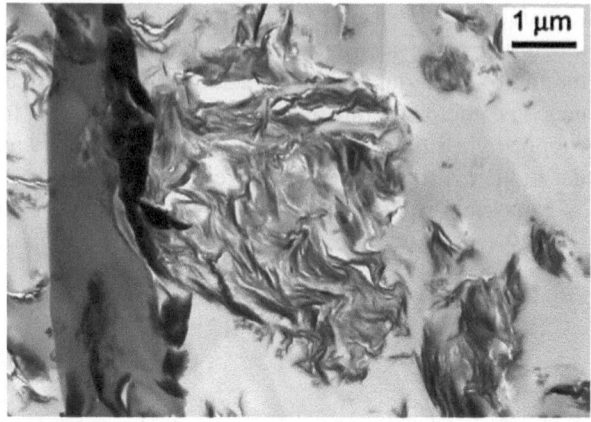

10–250 μm, the nanoscale reinforcement can exist in the interstitial spaces without compromising the packing factor of hollow particles. This arrangement is very important for syntactic foams because the weight reduction advantage is dependent on the particle packing factor in the composite. However, complete dispersion of nanoparticles, such as nanoclay, nanofibers, and graphene in polymeric resins is difficult to obtain and is expensive, which may increase the cost and processing time of the composite material. In addition, several studies have shown that nanoparticles stabilize gas bubbles in polymers [90–92], which may have an unintentional effect of increased matrix porosity if the processing method is not carefully designed.

The microstructure of CNF reinforced syntactic foam is shown in Fig. 3.4. Due to more than three orders of magnitude difference in the diameter of CNF and glass hollow particles, the nanofibers are not clearly visible at low magnifications. CNFs dispersed in the inter-particle region can be seen at higher magnifications. This arrangement allows strengthening the matrix without compromising the particle packing factor, which is important in reducing the density of the composite. In a similar manner, nanoclay has also been dispersed in the matrix in several studies [74, 93]. A CNF cluster is also seen in Fig. 3.4. The CNFs inside the cluster may not be wetted by the resin and existence of such clusters is expected to adversely affect the properties of the composite. Similar observations were also made in nanoclay reinforced syntactic foams. Figure 3.5 shows a transmission electron microscope (TEM) micrograph of nanoclay dispersion in epoxy matrix syntactic foams [94]. In one- and two-dimensional nanomaterials such as CNF and nanoclay, respectively, the surface area to volume ratio is very large, which allows formation of a large number of bonds between resin and reinforcement. Steep rise in viscosity of the resin is observed with only a small volume fraction, up to 5 wt% of nanomaterials due to their large surface area.

Chapter 4
Tensile Properties

Abstract Micro- or Nano-scale reinforcements are attractive for enhancing the tensile properties of syntactic foams. Glass and carbon fiber reinforcement was found to increase tensile strength and modulus of syntactic foams when compared to that of plain syntactic foam. The orientation of fibers with respect to the loading axis had a significant impact over the level of increase in the mechanical properties. However, reinforcement of microfibers increases the density of composite, which may be undesired in most applications; therefore, only a small volume fraction of fibers is used in syntactic foams. Carbon nanofiber (CNF) reinforcement increased the tensile strength and modulus of the syntactic foam. More specifically, addition of 0.25 wt% of CNFs increased, tensile modulus and strength of syntactic foams by 10–20 and 20–50 % (depending on the hollow particle wall thickness and volume fraction), respectively, when compared to that of plain syntactic foam. Nanoclay reinforcement was also found to increase the tensile strength and tensile modulus of syntactic foams in most studies. Nanofillers may have an undesired effect of stabilizing matrix porosity in syntactic foams during manufacturing. However, presence of nanoparticles around or across the matrix voids may provide a strengthening effect and the matrix porosity can be used to further decrease the syntactic foam density without a severe penalty on mechanical properties.

Keywords Syntactic foam • Hollow particle • Nanoscale material • Carbon nanofiber • Carbon nanotube • Nanoclay • Elastic modulus • Tensile strength • Tensile failure

The matrix is the only continuous phase in particulate reinforced composites. Therefore, it plays an important role in defining the tensile properties, especially the failure strength and failure mode. Widely used matrix materials such as epoxy and vinyl ester resins show brittle failure under tensile mode, which is also reflected in the failure of syntactic foams of these resins. Similar phenomenon is observed in reinforced syntactic foams, where matrix cracking is the main failure

N. Gupta et al., *Reinforced Polymer Matrix Syntactic Foams*, SpringerBriefs in Materials,   25
DOI: 10.1007/978-3-319-01243-8_4, © The Author(s) 2013

Table 4.1 Studies on tensile properties of reinforced syntactic foam

Reference	Matrix resin	Hollow particle material	Reinforcement type
[12]	Phenolic	Phenolic	Carbon fibers
[111]	Epoxy	Glass	Carbon nanofibers
[97]	Phenolic	Amino resin	Carbon and aramid fibers
[118]	Benzoxazine	Glass	Silica fibers
[119]	Phenolic	Poly-acrylonitrile	Polyester, glass, and aramid fibers
[120]	Epoxy	Phenolic	Short carbon fibers
[74]	Epoxy	Glass	Nanoclay
[121]	Cyanate ester	Glass	Nanoclay
[103]	Epoxy and PEEKMOH modified epoxy	Glass	Nanoclay

mode under tensile loading [95]. Table 4.1 summarizes the available studies on the tensile properties of reinforced syntactic foams.

Studies related to reinforced syntactic foams are mainly focused on using fibrous reinforcements to obtain effects such as crack bridging, in order to increase the energy absorption in the composite before failure and slowing down the crack growth rate. Incorporation of fibers may increase the failure strain of syntactic foams, but the failure is still seen in the brittle mode where cracks perpendicular to the loading direction cause the specimens to fracture at the end of a linear stress–strain curve [95, 96]. A detailed discussion on the tensile properties and failure mechanisms of reinforced syntactic foams is presented below.

4.1 Micro-Fiber Reinforced Syntactic Foams

In fiber reinforced composites, the fiber orientation is an important parameter that determines the composite properties. Effect of orienting fibers either parallel or perpendicular to the loading direction on the tensile properties of syntactic foams has been studied [97]. Tensile strength and modulus increased by 10–40 % when the fibers are oriented perpendicular to the loading direction, whereas an increase of 332–701 % is observed when the fibers are oriented parallel to the loading direction for different types and lengths of fiber. Generally it is easier to randomly disperse short fibers in composite materials. However, the reinforcing ability of fibers oriented in different direction is not the same. In such composites, effective fiber length $\left(\bar{l}\right)$ is calculated as a measure of effectiveness of fibrous reinforcement by

$$\bar{l} = \frac{N \int_0^{\pi/2} \int_0^{\pi/2} l \cos\theta \cos\phi d\theta d\phi}{N\left(\pi/2\right)^2} \tag{4.1}$$

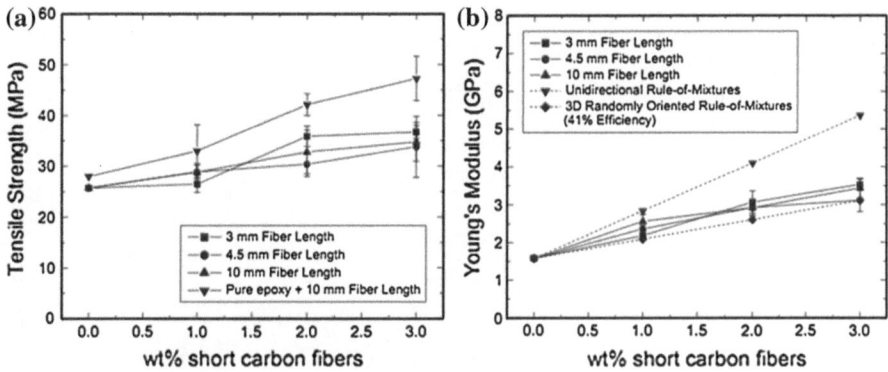

Fig. 4.1 a Tensile strength and **b** modulus of short carbon fiber reinforced syntactic foam with varying weight fractions and lengths of reinforcing fibers [96]

where l is the actual length of the fiber, N is the number of fibers, θ and ϕ are the angles projected on the loading axis, x, and x–y plane, respectively, in 3D orientation [96]. A randomly oriented three-dimensional discontinuous fiber reinforced syntactic foam showed that 41 % of the fiber length forms the effective length calculated by Eq. 4.1 [96].

The tensile test results on microfiber reinforced syntactic foams are shown in Fig. 4.1 [96]. The strength and modulus of syntactic foams increased almost linearly with the fiber content. In this work, the variation in fiber length did not show any measurable impact on the Young's modulus of reinforced syntactic foams because the short fibers used as reinforcement had lengths greater than the critical length needed for reinforcement [98]. Another work characterized syntactic foams reinforced with unidirectionally oriented carbon and aramid fibers of lengths of 12 and 24 mm, respectively [97]. The carbon fiber reinforcement perpendicular and parallel to the loading axis increased the tensile modulus of syntactic foam by 23 and 701 %, respectively. Similar configurations of aramid fibers increased the tensile modulus by 35 and 605 %, respectively. The tensile strength of syntactic foams containing carbon fibers parallel and perpendicular to the loading axis increased by 20 and 504 %, respectively and for aramid fiber reinforced syntactic foams by 387 and 709 %, respectively. These results show the possibility of improving the tensile strength of composites through microfiber reinforcement. Carbon and glass fibers have densities around 1800 and 2500 kg/m^3, respectively, which are considerably higher than that of polymer matrix and hollow particles. Addition of fibers increases the density of syntactic foams. Therefore, the increase in density should be justified in terms of strength enhancement.

4.2 Nanoscale Reinforced Syntactic Foams

Nanotubes, nanofibers, and nanoparticles have been extensively used in reinforcing polymers in recent years [56, 99–103]. Studies have shown that longer aspect ratio nanostructured reinforcements provide better mechanical properties

Fig. 4.2 TEM image of a
a functionalized MWCNT
in epoxy resin. **b** MWCNT
bridging a void in the matrix
[117]

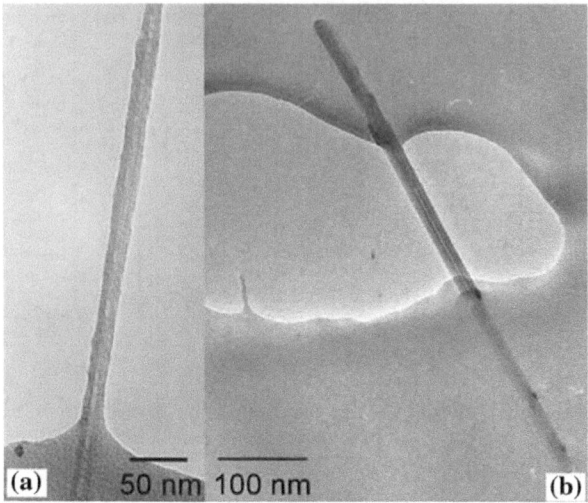

in composites [104]. Nanoscale reinforcement can help in effectively using the
matrix porosity as a beneficial microstructural feature. The matrix porosity can
help in reducing the syntactic foam density but deteriorates both modulus and
strength. Nanotubes and nanofibers can be used to reinforce the matrix voids and
limit the mechanical property reduction. An example of CNT bridging a void in
an epoxy matrix composite is shown in Fig. 4.2 [105]. Such reinforced voids can
reduce the density of the composite without a penalty on mechanical properties.

4.2.1 CNT and CNF Reinforced Syntactic Foams

CNTs and CNFs can enable mechanisms such as crack bridging and energy dissipation to improve properties of composites. However, difficulties in dispersing
large aspect ratio nanotubes and nanofibers have limited the mechanical property
enhancements obtained through these nanomaterials [106, 107]. Cost effectiveness of CNFs compared to nanotubes is leading to their increased applications
in composites [108–110], even in reinforcing syntactic foams [111]. A typical set of stress–strain curves for a 1 wt% CNF reinforced syntactic foam is presented in Fig. 4.3 [112]. The trend is similar to plain and micro-fiber reinforced
syntactic foams, where failure occurs at the end of the linear region. It is reported
that 10–20 % increase in tensile modulus can be obtained for 0.25 wt% addition
of CNFs [111]. In the same study, the tensile strength was found to increase by
20–50 % compared to plain syntactic foams of similar hollow particle volume fraction. Among the advantages of using CNFs in syntactic foams compared to glass
microfibers, are the possibilities of obtaining lower density syntactic foams, close
control over density of syntactic foams because of addition of a small volume

Fig. 4.3 Tensile stress-strain behavior of 1 wt% CNF reinforced epoxy matrix syntactic foams containing glass hollow particles (The nomenclature: N refers to nanofibers, followed by hollow particle density in kg/m^3 and then vol. %) [112]

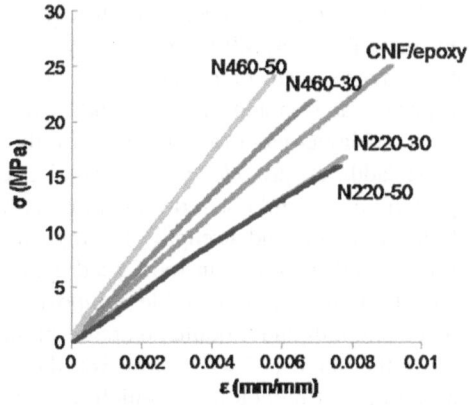

fraction of nanofibers, and no impact on the packing efficiency of hollow particles. Considerable future work is required to fully understand the mechanisms of reinforcements in CNF reinforced composites. The complex structure of CNF comprises a ribbon-like graphene structure that is helically folded to provide a fibrous structure with hollow core, which is termed as cup-stacked structure [113, 114]. Under tensile loading, the helically folded graphene sheet structure of CNFs may not provide reinforcement in the initial stages because it may have some freedom to unfold. The bonding between various folded layers of graphene determines the strength and stiffness of such nanofibers [114].

4.2.2 Nanoclay Reinforced Syntactic Foams

Nanoclay has been used in reinforcing epoxy matrix syntactic foams [74, 103, 115]. Extensive use of nanoclay in reinforcing various types of polymers in recent years has resulted in development of processing methods that can exfoliate and disperse nanoclay [33, 116]. The stress–strain curves showed a linear trend until fracture in nanoclay reinforced syntactic foams. Compared to the plain syntactic foams, composites containing 2 and 5 vol. % nanoclay showed up to 23 % increase in tensile strength for foams containing 63 and 60 vol. % glass hollow particles of several wall thicknesses [74]. On the contrary, the Young's modulus of most reinforced syntactic foams was lower than the plain syntactic foams. In these foams, complete exfoliation of nanoclay was not obtained and clusters were present. The intercalated and exfoliated parts of the clay may have provided some toughening but the relatively low modulus of nanoclay clusters, due to the possibility of platelet sliding with respect to each other, contributed to overall reduction in the modulus. The effect of presence of clusters also resulted in large standard deviation in the experimental measurements of tensile strength and modulus. A similar work on cyanate ester syntactic foams showed increase of about 60 and 90 % in tensile strength for 2 and 4 vol. % nanoclay reinforced syntactic foams. The tensile modulus also increased by 6 and 80 % in these syntactic foams. Better exfoliation of nanoclay in the matrix resin obtained in this work resulted in improvement of mechanical properties. It is noted in some studies on clay/epoxy nanocomposites that at certain nanoclay content (around 2 wt%), the mechanical properties show a peak and further addition provides a detrimental effect [115]. However, nanoclay reinforced syntactic foams did not show such trends within the range of compositions studied. Higher modulus and strength of composites were observed with increasing hollow particle wall thickness in reinforced syntactic foams, similar to those of the plain syntactic foams of comparable compositions. It is also observed in a study that the tensile strength and modulus increase monotonically with increasing nanoclay content in the range 1–5 wt%. However, the trend in the specific modulus and specific strength is not monotonous, which indicates that the matrix porosity may not be in the close range in the fabricated composites and is an important factor. The matrix used in these syntactic foams is an epoxy resin modified with hydroxyl terminated polyether ether ketone having pendant methyl group (PEEKMOH). The tensile modulus of this toughened resin filled with 40 wt% hollow particles is measured as 1.29 GPa. The nanoclay reinforcement results in the highest modulus value of 1.5 GPa at 5 wt% nanoclay dispersion.

4.3 Remarks

Ideally, the studies on nanoclay reinforced syntactic foams should provide mechanical properties of neat polymer, plain syntactic foams, and nanoclay reinforced syntactic foams to facilitate correlation among their properties and draw clear conclusions. However, most studies have not provided such comparisons. In addition, data obtained from different sources can be compared, but differences in the hardener type and volume fraction, cure cycle, and specimen storage conditions can reflect in mechanical properties and contaminate the results. Therefore, there is still ample scope of systematic future experimental investigations that can provide clear comparison between plain and nanoscale reinforced syntactic foams.

Chapter 5
Modeling and Simulation

Abstract Development of theoretical models is very important for syntactic foams. Numerous parameters are involved in syntactic foam design, which include matrix and particle material, particle volume fraction and wall thickness, and reinforcement material and volume fraction. To identify the parameters that would result in syntactic foam with desired set of mechanical and thermal properties, theoretical models can be very useful and cut down the need for experimentation. Several existing models applicable to particulate composites can be modified to include the particle wall thickness effect. Multiscale models that can include the nanofibers or particles along with hollow particles are not available yet. It is challenging to model syntactic foams that contain high volume fractions of hollow particle (close to packing limit) because of particle-to-particle interaction effects. Two models are used in the present chapter to estimate the properties of multiscale syntactic foams. Both models are applicable to plain syntactic foams containing only hollow particles in matrix. Therefore, semi-empirical approach is adopted and the experimentally measured properties of nanofibers reinforced polymer are assigned to the matrix. The model predictions are validated with experimental results. Finite element analysis is especially illustrative in understanding the behavior of syntactic foams under the applied load. A validated finite element study conducted on a unit cell geometry comprising a hollow particle and a fiber showed that the particle wall thickness plays an important role in determining the stress distribution in microscale reinforced syntactic foam system. For syntactic foams containing thin-walled particles, the location of the maximum von-Mises stress exists within the particle, whereas above a critical wall thickness the location of the maximum stress shifts to the fiber. This pattern illustrates that the location of failure initiation can be tailored in syntactic foams.

Keywords Syntactic foam • Hollow particle • Nanoscale material • Carbon nanofiber • Carbon nanotube • Nanoclay • Elastic modulus • Homogenization technique • Differential scheme • Finite element analysis • Multiscale modeling • Density-strength relation • Density-modulus relation

N. Gupta et al., *Reinforced Polymer Matrix Syntactic Foams*, SpringerBriefs in Materials, 31
DOI: 10.1007/978-3-319-01243-8_5, © The Author(s) 2013

The reinforced syntactic foams that are experimentally studied include reinforcements of a wide variety size scales, shapes, and properties. Any one approach may not be able to model the behavior of the entire range of reinforced syntactic foams. This chapter summarizes various modeling and simulation approaches available for reinforced syntactic foams and presented results on syntactic foam systems that have been experimentally studied.

5.1 Microscale Reinforced Syntactic Foams

Several modeling approaches are available for estimating the elastic properties of syntactic foams [122–125]. While theoretical models are not available on fiber reinforced syntactic foams, insight into the properties of such composites can be obtained from a simulation study. Finite element analysis is conducted on a unit cell geometry comprising a hollow particle and a fiber in this study [126]. The calculated elastic modulus from this model matched closely with the experimental results taken from the literature [96]. Results showed that the particle wall thickness plays an important role in

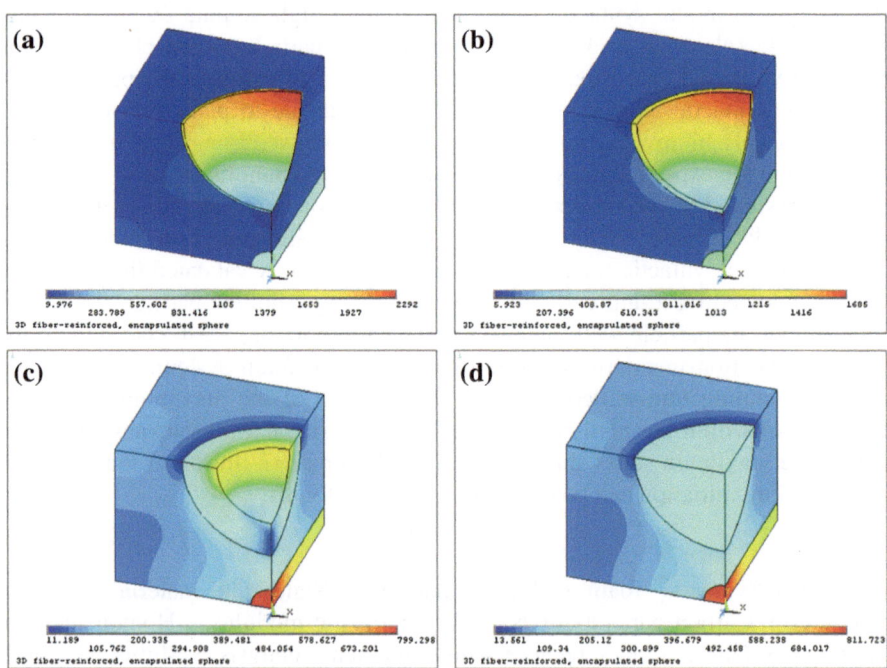

Fig. 5.1 Finite element analysis results of von-Mises stress for *y*-direction loading of a unit cell model of a fiber reinforced syntactic foam. For all other parameters maintained constant, the effect of wall thickness on the stress distribution can be observed. The particle wall thickness is increasing sequentially from **a** to **d**. The location of maximum stress is found to be in the hollow particle or the fiber for syntactic foams containing thin or thick-walled hollow particles, respectively

determining the stress distribution in the material system as shown in Fig. 5.1 [126]. For syntactic foams containing thin-walled particles, the location of the maximum von-Mises stress exists within the particle, whereas above a critical wall thickness the location of the maximum stress shifts to the fiber. These observations are important in determining the parameters that would provide the desired energy absorption and failure characteristics in reinforced syntactic foams. A combination of particle wall thickness and volume fraction can be used to tailor the failure mode and energy absorption characteristics of syntactic foams. This analysis considered dilute dispersion of particles in the matrix to neglect particle-to-particle interactions and focused only on the particle-to-fiber interactions. Understanding the interaction between particles of different size and wall thicknesses will be interesting in future studies [127].

Recent literature has extensively studied particle-to-particle interaction effects in syntactic foams [128, 129] and extension of these studies to cover reinforced syntactic foams is yet to be conducted. Stress profiles, stress intensification factors, circumferential stress resultants in the particles, and strain energy storage are analyzed in these studies to understand the interaction effects and derive the effective elastic properties of syntactic foams. In general, it is observed that neglecting the particle-to-particle interaction effects overestimates the stiffness of syntactic foams containing thin-walled particles and underestimates for foams containing thick-walled particles [128, 129]. The critical wall thickness that separates these two types of behaviors is a function of the ratio of the stiffness of particle material and matrix resin. These analyses should be extended to include the effect of fiber volume fraction, length, and diameter on stress profiles and the effective elastic properties of reinforced composites.

5.2 Nanoscale Reinforced Syntactic Foams

Numerous theoretical models are available to predict the elastic modulus of particle reinforced composites [130–132]. Some of the existing models are specialized for syntactic foams to account for the hollow particle wall thickness as a parameter [124, 125, 132]. Modeling approaches applicable to syntactic foams have been discussed in a focused article to demonstrate their application domains, strengths, and limitations [133]. Classical composite sphere based self-consistent schemes have been extended to plain syntactic foams [125]. In addition, several three- or four-phase models are now available that can be applied to predict the elastic properties of syntactic foams [134, 135].

5.2.1 Porfiri-Gupta Model

Classical solutions for dilute dispersion of solid particles in an infinite matrix can be adapted to estimate elastic properties of syntactic foams containing small volume fractions of hollow particle [130]. However, most syntactic foams contain high volume fractions of hollow particle (mostly approaching the packing limit) to obtain the maximum benefit of the weight saving potential. Analysis of such composites is challenging because of particle-to-particle interaction effects. Recently, a differential scheme is used to estimate the elastic properties of syntactic foams containing high volume fraction of hollow particles [31, 132]. This scheme is based on homogenization of dilute dispersion of particles in an infinite matrix and is iteratively used to obtain high particle volume fraction as shown in Fig. 5.2. In the first step, a dilatation and a shear problem are solved to compute the elastic properties of the composite containing dilute dispersion of particles [132, 136]. In the next step, a differential scheme is used to extrapolate the properties of dilute dispersion of composites containing high volume fractions of particles. This scheme results in a coupled nonlinear differential equation set

$$\frac{dE}{E} = f_E(E_b, v_b, E_m, v_m, \eta) \frac{d\Phi}{\left(1 - \Phi/d\Phi\right)} \qquad (5.4a)$$

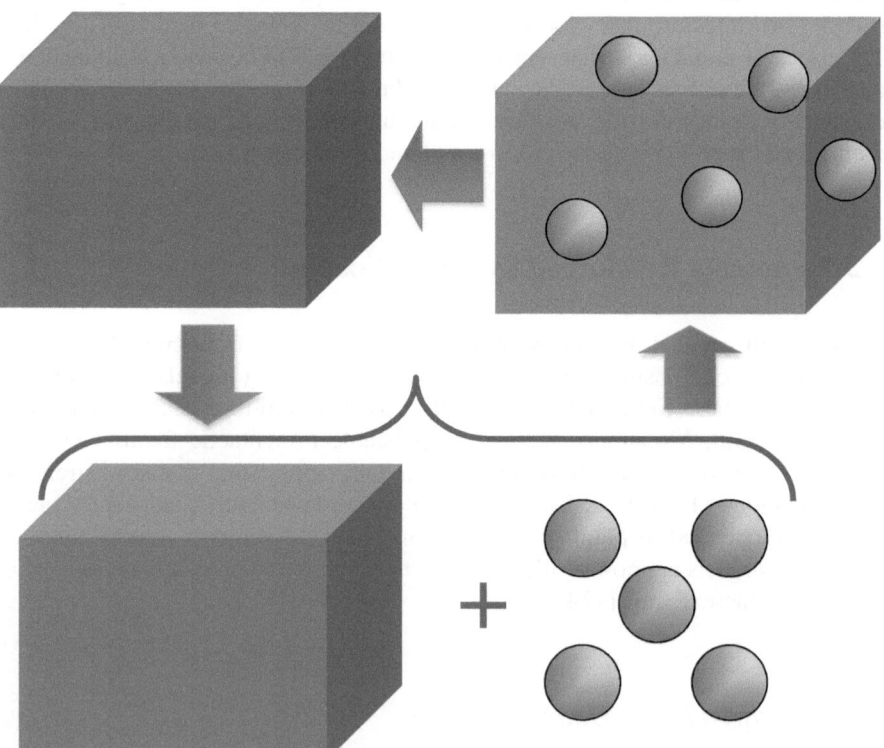

Fig. 5.2 Schematic representation of differential scheme where a dilute dispersion solution is iteratively used to estimate the elastic properties of syntactic foams containing high volume fraction of hollow particles

$$\frac{dv}{v} = f_v(E_b, v_b, E_m, v_m, \eta) \frac{d\Phi}{(1 - \Phi/d\Phi)} \qquad (5.4b)$$

Where E_b and v_b are elastic modulus and Poisson's ratio of hollow particle material and E_m and v_m are elastic modulus and Poisson's ratio of matrix material, respectively. In addition, Φ is the hollow particle volume fraction. Equation set 5.4 is solved to obtain the elastic properties of syntactic foams. Detailed expressions for equation set 5.4 are found in [31]. The differential scheme is also useful in incorporating the effect of polydispersivity in particle size and wall thickness [127]. Particles of different geometrical properties can be incorporated in the composite during different iterations to obtain more realistic predictions of elastic properties of syntactic foams [127].

The differential scheme can be used to estimate the elastic properties of nanoparticle reinforced syntactic foams. In general, the size of nanoparticles is three orders of magnitude smaller than the commonly used hollow particles. The nanoparticle reinforced resin can be taken as the homogenized media in estimating the properties of multiscale reinforced syntactic foams. The applicability of this modeling scheme for nanoscale reinforced syntactic foams is shown in Fig. 5.3, where experimental and theoretical values of Young's modulus for a variety of CNF and nanoclay reinforced syntactic foams are found to be in close agreement. The Young's moduli used in the theoretical calculations are 3.2 and 1.4 GPa for the CNF and nanoclay reinforced resins, respectively, and the Poisson's ratio for both materials is taken as $v_m = 0.3$. The properties of the hollow particle material are taken as $E_b = 60$ GPa and $v_b = 0.21$, respectively [137]. For the data displayed in Fig. 5.3b [74], the nanoclay reinforced syntactic foams contain 60 vol. % hollow particles. These results show that the differential scheme can be used for

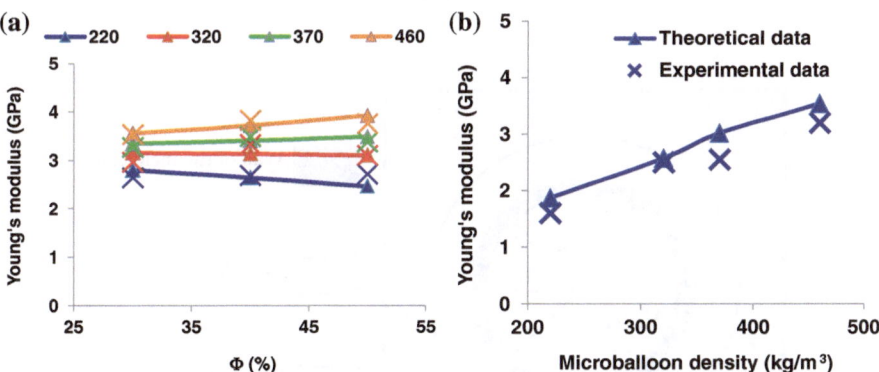

Fig. 5.3 Comparison of experimental values with the predictions obtained from the differential scheme for **a** 0.25 wt% CNF reinforced syntactic foams, experimental data are taken from [111]. The experimental data are represented by symbol X of appropriate color and the legend represents the hollow particle density in kg/m^3 and **b** 5 vol. % nanoclay reinforced syntactic foams, experimental data are taken from [74]

estimating the properties of multiscale reinforced syntactic foams. The modulus of syntactic foams decreases with increasing volume fraction of thin-walled hollow particles, but the trend reversed for thick-walled particles. The results show that the composites having a wide range of mechanical behavior can be engineered by using nanoscale reinforcements.

5.2.2 Bardella-Genna model

Bardella and Genna proposed a four-phase model (Fig. 5.4) to estimate the elastic properties of syntactic foams [125]. The equivalence of the average strain over the entire hollow particle is utilized to predict the composites shear (μ) and the bulk (k) moduli [125]. The applied strain field E_{12} at the boundary of the representative volume element is concurred with the volume average of the corresponding local fields in the considered four-phase composite model and is represented as

$$E_{12} = \langle \gamma_{12} \rangle \tag{5.6}$$

where γ_{12} is the shear strain. The average stress–strain relationship for the representative volume element can be written as

$$2\mu \langle \gamma_{12} \rangle = \langle \tau_{12} \rangle \tag{5.7}$$

For the presence of N different composites spheres, the shear strain can be expressed in terms of the volume average shear strain computed on the single composites sphere as

$$\langle \gamma_{12} \rangle = \sum_{j=1}^{N} \chi_j \overline{\gamma}_{12}^{C.S.} \tag{5.8}$$

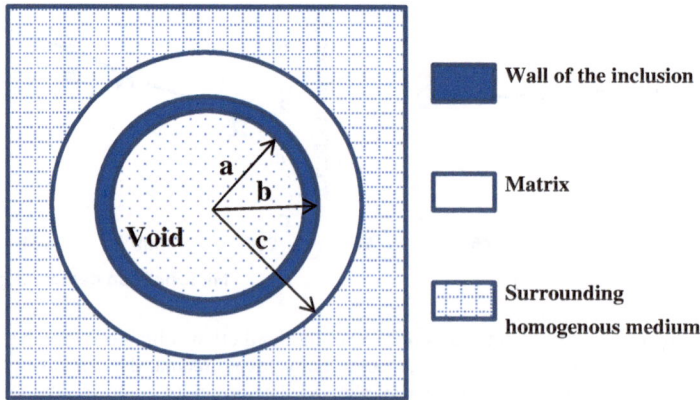

Fig. 5.4 The four-phase model utilized in Bardella-Genna model [125]

where the superscript *c.s.* represents composite sphere and χ_j represents the fraction of the filler type j to the entire filler. The volume average stress is expressed as

$$\langle \tau_{12} \rangle = 2 \sum_{j=1}^{N} [\mu_{i,j} \chi_{i,j} \overline{\gamma}_{12}^{i,j.} + \mu_{m,j} \chi_{m,j} \overline{\gamma}_{12}^{m,j.}] \tag{5.9}$$

where $\chi_{m,j}$ and $\chi_{i,j}$ represent the volume fraction of the matrix and the inclusion of the composites sphere j, respectively. Substituting Eqs. (5.8) and (5.9) into Eq. (5.7), the homogenized shear modulus can be written as

$$\mu = \frac{\displaystyle\sum_{j-1}^{N} \chi_j \left\{ \mu_{i,j} \Phi \left[1 - \left(\frac{r_i}{r_o} \right)^3 \right] \overline{\gamma}_{12}^{i,j.} + \mu_{m,j} (1 - \Phi) \overline{\gamma}_{12}^{m,j.} \right\}}{\displaystyle\sum_{j=1}^{N} \chi_j \overline{\gamma}_{12}^{(C.S.),j.}} \tag{5.10}$$

where i and m are the inclusion and the matrix shell and Φ is the volume fraction of the filler in the syntactic foam. The bulk modulus is also obtained in a similar fashion as the shear modulus and is expressed as

$$k = \frac{\displaystyle\sum_{j-1}^{N} \chi_j \left\{ k_{i,j} \Phi \left[1 - \left(\frac{r_i}{r_o} \right)^3 \right] \overline{\varepsilon}_{kk}^{i,j} + k_{m,j} (1 - \Phi) \overline{\varepsilon}_{kk}^{m,j} \right\}}{\displaystyle\sum_{j=1}^{N} \chi_j \overline{\varepsilon}_{kk}^{(C.S.),j}} \tag{5.11}$$

where ε_{kk}, γ_{12} are average volumetric strain and shear strain, respectively. The detailed derivation of these expression and the related coefficients are given in [125]. The values obtained from the Bardella-Genna model are compared with the experimental results for CNF reinforced and nanoclay reinforced syntactic foams and are represented in Fig. 5.5. The modulus value of the homogenized reinforced matrix is taken the same value as in the differential scheme, to get comparable estimates. The predictions from this model are lower than the experimental values. However, the difference between experimental and theoretical values is small. In general, the predictions obtained from this model are lower than the values obtained from the Porfiri-Gupta model.

5.2.3 Parametric Study

Further analysis based on the differential scheme is presented in Fig 5.6 for two widely used particle types using Porfiri-Gupta model. In the first step, the modulus of neat epoxy is taken as 2.75 GPa [5] to calculate the modulus of CNT/epoxy nanocomposite (σ_c) using the relation reported in [138].

$$\sigma_C = \sigma_m + 0.07 \varphi_{CNT} \tag{5.5}$$

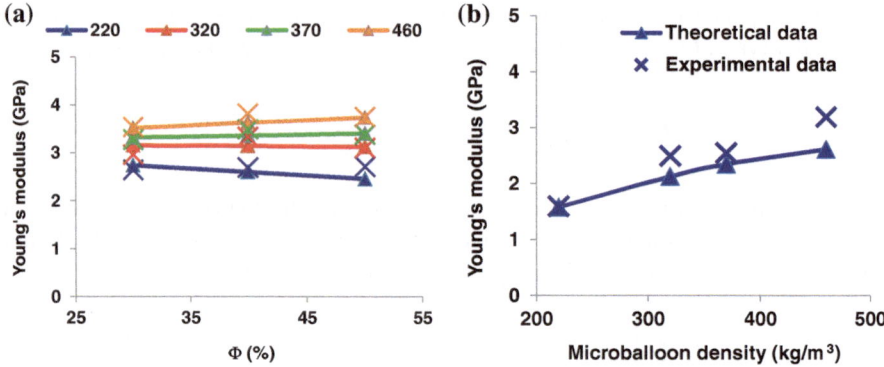

Fig. 5.5 Comparison of experimental values with the predictions obtained from the Bardella-Genna model for **a** 0.25 wt% CNF reinforced syntactic foams, experimental data are taken from [111]. The experimental data are represented by symbol X of appropriate color and the legend represents the hollow particle density in kg/m³ and **b** 5 vol. % nanoclay reinforced syntactic foams, experimental data are taken from [74]

Fig. 5.6 Predictions of change in Young's modulus of CNT reinforced syntactic foams for 1–10 wt.% CNTs and 30–60 vol.% hollow particles of 220–460 kg/m³ density. The modulus values of CNT reinforced epoxy are derived on the basis of trend reported in [138]. The elastic modulus of neat epoxy resin is taken as 2.75 GPa [5]

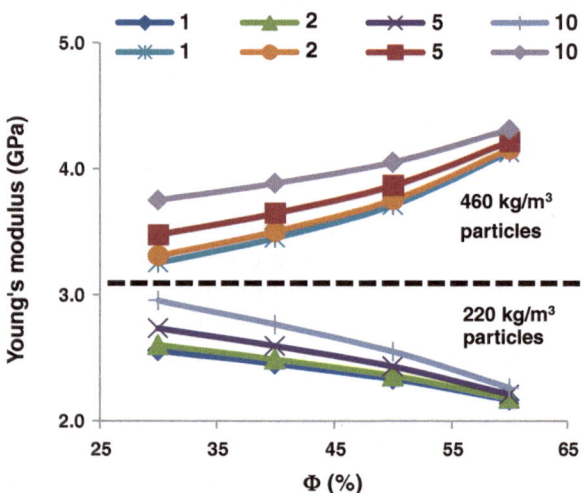

where σ_m and ϕ_{CNT} are the modulus of neat matrix resin and wt% of CNTs. In the second step, the differential scheme is applied to calculate the modulus of reinforced syntactic foams. A thin-walled and a thick-walled particle types are selected for plotting the modulus values in Fig. 5.6. It can be observed that at low volume fractions of hollow particles, the modulus of multiscale composite has a strong dependence on CNT content. However, the sensitivity of modulus to the CNT content decreases as the matrix volume fraction is reduced in the composite.

In the present case, the experimentally measured properties of nanoparticle reinforced resins or trends based on empirical measurements are used as input parameters in the differential scheme. Further development in the predictive models can include obtaining theoretical estimates for the properties of CNF or nanoclay reinforced resins that can be used in differential schemes. Experimental validation of theoretical results is also required for various material types.

5.2.4 Molecular Simulation Methods

Molecular simulations, specifically molecular dynamic (MD) simulations, have been recently used for prediction of elastic properties of nanoscale filler rein-forced composite materials [139, 140]. CNTs are widely studied using molecular simulation methods [70, 141, 142]. MD simulations are also used to study the influence of chemical bonding between a single-walled CNT and a polymer (crystalline or amorphous polyethylene) matrix on the shear strength of CNT-matrix interface [143]. A unit cell that is used in a particular MD simulation containing several randomly oriented CNTs in a box filled with epoxy molecules is shown in Fig. 5.7. The existing studies are mainly focused on a single CNT present in the system. It is shown that the shear strength of the CNT-polymer interface with weak nonbonded interactions can be increased by over an order of magnitude with the introduction of a relatively low density (<1 %) of chemical bonds between the CNT and matrix. MD simulations are also used to analyze the stress–strain behavior of single-walled CNT-polymer composites mechanically loaded in longitudinal or transverse directions [144]. In this study, the CNT aspect ratio played an important role and long nanotubes provided a stiffening effect in the composite while short nanotubes provided no enhancement in stiffness compared to the neat resin. Computational efforts exploring different types of nanotubes and their influence on electrical, mechanical, and thermal properties are discussed in a review article [27]. The mechanical properties of nanocomposites estimated by such schemes can be used as inputs in differential scheme to incorporate the effect of hollow particles and build multiscale model for nanoscale reinforced

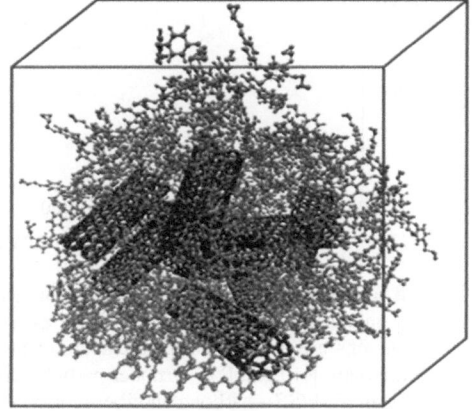

Fig. 5.7 A unit cell filled with epoxy resin containing several randomly oriented CNTs. Such models are used for molecular dynamic simulations

syntactic foams. Such an approach will eliminate the requirement of using empirical properties of nanoscale reinforced resins into the differential scheme and other models used for syntactic foams.

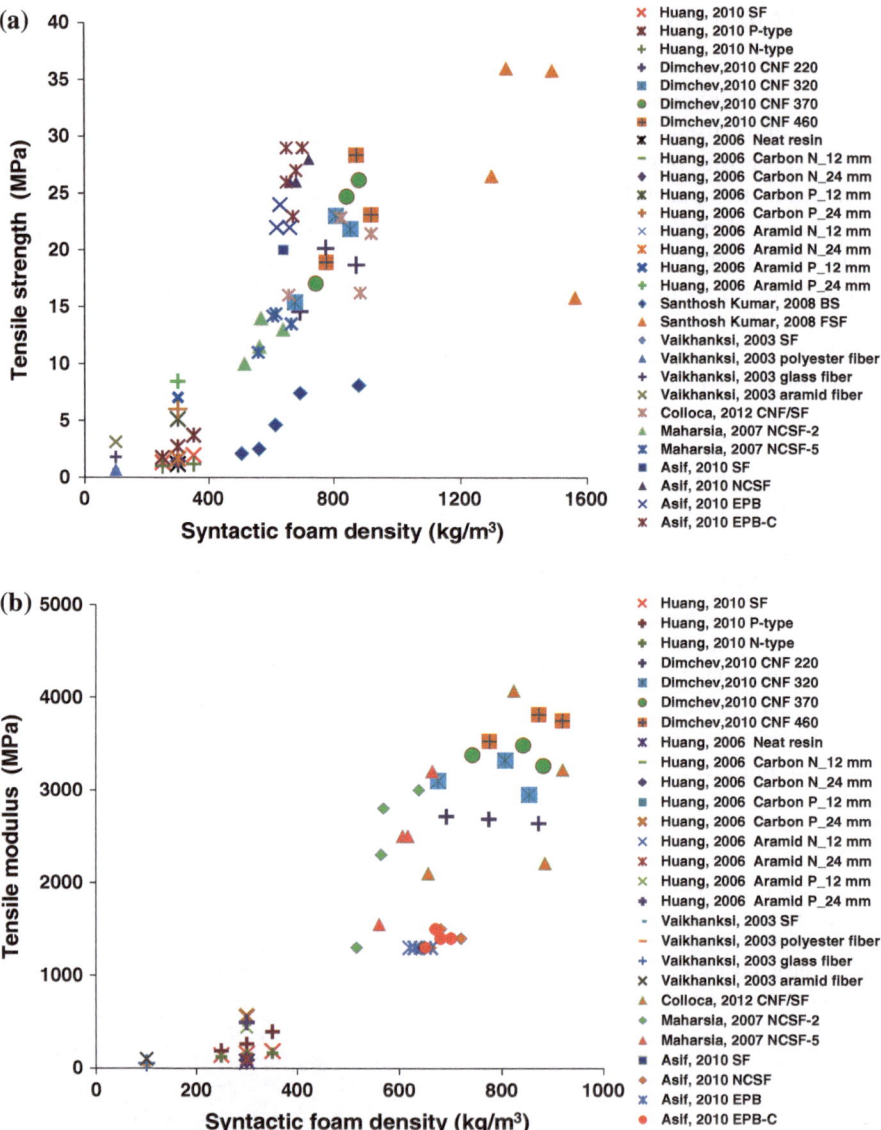

Fig. 5.8 a Tensile strength and **b** modulus of reinforced syntactic foam plotted against the syntactic foam density. Only average values are extracted for the respective properties and the standard deviations are not plotted. Nomenclature: SF—plain syntactic foam; FSF—fiber reinforced syntactic foam, R40 and R75—rubber particle reinforced syntactic foam; NCSF—nanoclay syntactic foam; CNF/SF—carbon nanofiber reinforced syntactic foam

5.3 Relation Between Density and Tensile Properties

Table 4.1 shows that a wide variety of matrix, particle, and reinforcement materials are used in the available studies. This makes it difficult to directly compare the mechanical properties reported in various sources. Since the low density of syntactic foams is the most promising feature that enables their applications, the trends in mechanical properties are analyzed with respect to density as shown in Fig. 5.8. Tensile strength and modulus data are obtained from published articles and are plotted with respect to density in this figure to identify general trends. Some of the data are extracted from figures with the best possible accuracy. It should be noted that any reduction in the density and mechanical properties due to matrix porosity is inherently included in the data presented in Fig. 5.8.

It can be observed in Fig. 5.8a that higher density foams have higher tensile strength and the trend is generally linear for the entire data set. The slope calculated based on the broadly linear trend of the entire data set, excluding the outlying points of [118], is 0.03 MPa/kg/m^3. Only the tensile strength data related to fiber reinforced benzoxazine resin matrix syntactic foams fall out of the approximately linear trend. Figure 5.8b shows that up to a syntactic foam density of about 700 kg/m^3, the modulus trend can be approximated to be linear with a slope of 4.72 MPa/kg/m^3. Increasing the density above around 700 kg/m^3 does not lead to further improvement in the tensile modulus of reinforced syntactic foams. Generally, syntactic foams containing high density hollow particles (true particle density of 460 kg/m^3 and above) are located in the high density range.

Figure 5.8 shows that with the choice of appropriate constituent materials and concentrations, the tensile strength and modulus can be tailored over a wide range. The tensile strength and modulus are found to be as high as 30 MPa and 4.2 GPa, respectively, in reinforced syntactic foams. It is desired to have higher mechanical properties for lower densities, but it appears from these graphs that the potential for weight saving is within a narrow range by changing the material composition. It is desired to find ways to increase the strength and modulus of low density syntactic foam compositions to further increase the weight saving potential in engineering applications. Development of new polymeric resins that have the same density but higher modulus and strength can help in achieving this goal.

5.3 Relation Between Ductility and Tensile Properties

Chapter 6
Compressive Properties

Abstract Most existing applications of syntactic foams are based on their compressive properties. Hollow particles are load bearing elements in the syntactic foam microstructure under compression, which helps in obtaining a long stress plateau region in the stress–strain graphs that helps in obtaining high energy absorption. The available studies have extensively studied the effect of hollow particle wall thickness and volume fraction on compressive properties of syntactic foams. Similarly, reinforced syntactic foams have been extensively studied for compressive properties. Carbon nanofibers and nanotubes, nanoclay, crumb rubber and glass, and carbon fibers have been used as reinforcements in syntactic foams to tailor the compressive properties. The incorporation of CNFs increases the overall energy absorption capacity of the composite. Orientation of fibers with respect to the loading axis is important in obtaining strengthening effect and randomly dispersed fibers do not provide high level of increase in compressive properties. In some layered reinforced syntactic foams, fibers oriented parallel to loading direction enhanced their compressive strength by 180–220 % compared to that of plain syntactic foams. On the contrary, foams containing fiber orientation perpendicular to the loading direction showed no measurable change in strength. The strength and modulus increase with density for reinforced syntactic foams. The maximum compressive strength and modulus for reinforced syntactic foam were found to be 120 MPa and 2.2 GPa, respectively.

Keywords Syntactic foam • Hollow particle • Nanoscale reinforcement • Carbon nanofiber • Carbon nanotube • Nanoclay • Compressive modulus • Compressive strength • Particle crushing • Density-strength relation • Density-modulus relation

A large number of syntactic foam applications are based on their compressive properties. The compressive stress–strain graphs of syntactic foams show a stress plateau region for large strain [145, 146], up to 50 % for some compositions, providing high energy absorption capabilities to these materials. The compressive

N. Gupta et al., *Reinforced Polymer Matrix Syntactic Foams*, SpringerBriefs in Materials, 43
DOI: 10.1007/978-3-319-01243-8_6, © The Author(s) 2013

Table 6.1 Studies on compressive properties of reinforced syntactic foams

Reference	Matrix resin	Hollow particle material	Reinforcement type
[12]	Phenolic	Phenolic	Carbon fibers
[157]	Epoxy	Glass	Glass fibers
[89]	Epoxy	Glass	Glass fibers
[111]	Epoxy	Glass	Carbon nanofibers
[97]	Phenolic	Amino resin	Carbon and aramid fibers
[155]	Epoxy	Glass	Glass fibers
[158]	Epoxy	Glass	Glass fibers
[118]	Benzoxazine	Glass	Silica fibers
[119]	Phenolic	Poly-acrylonitrile	Polyester, glass, and aramid fibers
[149]	Epoxy	Glass	Nanoclay
[121]	Cyanate ester	Glass	Nanoclay
[103]	Epoxy modified with PEEKMOH	Glass	Nanoclay
[150]	Phenolic-epoxy	Glass	Carbon nanotubes
[147]	Epoxy	Glass	Crumb rubber

stress–strain graphs of syntactic foams can be divided into three regions: an initial linear elastic region, a stress plateau region where stress remains nearly constant, and a densification region where stress starts rising again [145]. Usually these characteristics are retained by reinforced syntactic foams. The addition of second phase particles can help in tailoring the modulus, plateau stress, and densification strain and provide characteristics that are suitable for a given application.

Crushing of hollow particles is an important failure mechanism under compression. Therefore, volume fraction, wall thickness, and material type of particles have been widely studied for their effect on compressive properties of plain syntactic foams. Similar effects have also been characterized for several types of reinforced syntactic foams, as summarized in Table 6.1. Glass, aramid, silica, and carbon fibers are used for enhancement of compressive properties. In addition, nanoscale reinforcement by CNF and nanoclay is explored as indicated in Table 6.1.

6.1 Compressive Stress–Strain Behavior

The compressive stress–strain graphs of plain and CNF reinforced syntactic foams are compared in Fig. 6.1a. Randomly dispersed CNFs provide a reinforcing effect to the resin, including increases in modulus and strength. The characteristic stress-plateau region appears in this figure for both plain and CNF reinforced syntactic foams. The overall energy absorption capacity of the composite increases due to the incorporation of CNFs. The stress–strain behavior and failure mechanisms

Fig. 6.1 Stress–strain
graph of **a** an unreinforced
(VE460-30) and 1 wt% CNF
reinforced (N460-30) vinyl
ester (VE) matrix syntactic
foams containing 30 vol. %
glass hollow particles of
460 kg/m^3 density and
b syntactic foams containing
5 vol. % nanoclay with
60 vol. % glass hollow
particles of three different
true particle densities of 220,
320, and 460 kg/m^3

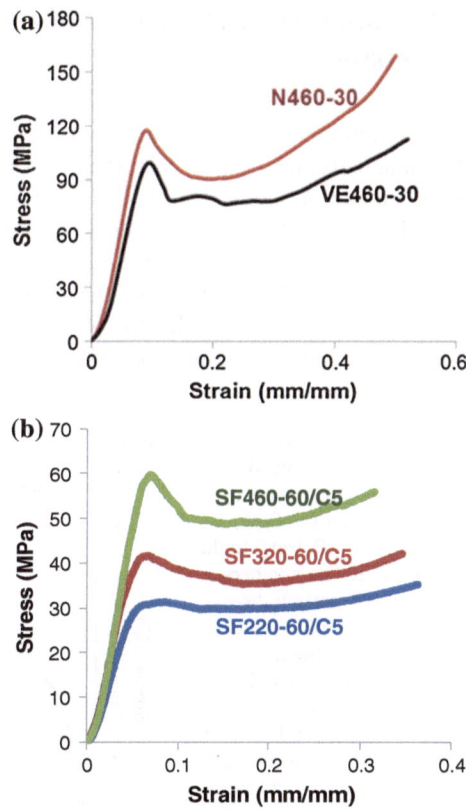

depend on the volume fraction and orientation of the reinforcing phase in the composite. A strong influence of hollow particle wall thickness is observed on compressive properties of nanoclay reinforced syntactic foams in Fig. 6.1b, because the
particles are the load-bearing component in the system under compression. Thicker
walled particles provided higher modulus and strength to syntactic foams at the
same reinforcement volume fraction level. Figure 6.1 shows a large compressive
strain in the specimens before densification. The length (densification strain) and
height (plateau stress) of the stress plateau are tailored by means of second phase
reinforcements in the available studies. Use of compliant particles in the matrix can
result in reduction in plateau strength but increase in densification strain. On the
contrary, use of stiffer reinforcement can lead to higher strength but it may com-
promise densification strain. The area under the stress–strain curves up to the densification strain is taken as a measure of energy absorption capacity of syntactic
foams in several studies. Many applications have complex requirements that go
beyond elastic constants and include parameters such as thermal expansion, moisture absorption, high temperature stability, and glass transition temperature that can
also be controlled through matrix modification using second phase particles.

6.2 Compressive Properties of Fiber Reinforced Syntactic Foams

The compressive properties are dependent on the orientation of reinforcing fibers in syntactic foams [12, 89, 97]. Higher compressive strength values for foams with fiber orientation along loading direction were observed [89]. The specimen failure also showed different trends when the fibers were oriented parallel and perpendicular to the loading direction. Extensive hollow particle crushing resulting in debris formation was observed due to the limited load bearing capacity of fibers when they are oriented perpendicular to the loading direction [12, 89]. On the other hand, when the fibers are oriented parallel to the loading direction, a relatively

Fig. 6.2 Compressive stress–strain graphs of syntactic foams with reinforcement **a** perpendicular (N-Type) and **b** parallel (P-Type) to the loading axis. The numbers in the plots represent the following type of syntactic foams: (1) neat foam, (2) N-type short aramid fiber-reinforced foam, (3) N-type short carbon fiber-reinforced foam, (4) N-type long aramid fiber-reinforced foam, (5) N-type long carbon fiber-reinforced foam, (6) P-type short aramid fiber-reinforced foam, (7) P-type short carbon fiber-reinforced foam, (8) P-type long carbon fiber reinforced foam, and (9) P-type long aramid fiber-reinforced foam. The figure is adapted from [97]

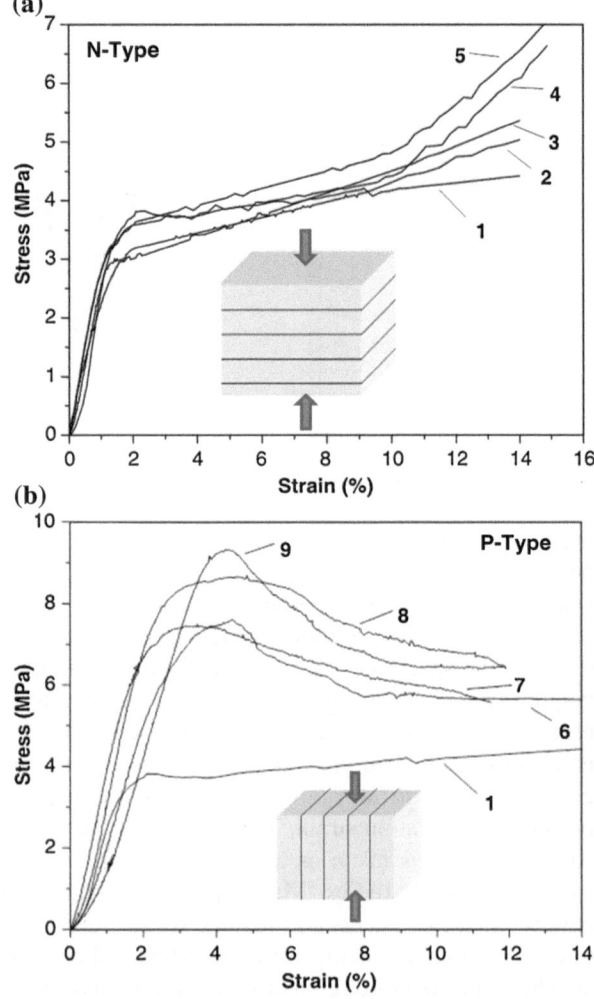

large number of uncrushed hollow particles are seen in the specimens and the fibers played a greater role in improving the compressive properties of syntactic foams.

Compared to the plain syntactic foams, the strength values of reinforced syntactic foams are found to be 180–220 % higher when the fibers are oriented along the loading direction in some layered syntactic foams, as shown in Fig. 6.2 [97]. On the contrary, foams containing fiber orientation perpendicular to the loading direction showed no measurable change in strength. The compressive modulus was largely unaffected by the aramid fiber orientation, but an increase of 40 % was reported for foams reinforced with carbon fibers along the loading direction [12].

The compressive properties of reinforced syntactic foams increase with density. Despite enhancing mechanical properties, glass fibers can lead to increased density of syntactic foams, which can limit their weight saving capability. It can also be concluded that controlled orientation of fibers can provide a desired level of compressive property enhancement in syntactic foams. The interaction between fibers and particles can have significant effects in microfiber reinforced syntactic foams because their size is of the same order. Detailed modeling and analytical studies can improve the understanding of such materials but such information is not yet found in the published literature and can be a focus of future work.

6.3 Compressive Properties of Rubber Reinforced Syntactic Foams

Applications of crumb rubber particles in toughening of matrix resins can be found. Mechanically ground waste tire rubber particles of 40 and 75 μm diameter are dispersed in several types of syntactic foams [147, 148]. When the same volume fraction of crumb rubber and hollow glass particles are used in the composite, no measurable effect of rubber particle size was found on the compressive strength and modulus of syntactic foams [147]. The rubber toughened foams showed only a small decrease in compressive strength, however, the modulus decreased by about 50 % in most compositions compared to plain syntactic foams. The stability of the stress plateau increased with the addition of rubber particles. It is observed that the rubber toughened syntactic foams had a stress-plateau extended to about 40 % compressive strain. A combination of modulus, strength, and densification strain can help in developing syntactic foam compositions that are tailored for the requirement of a given application.

Other configurations of rubber toughening of syntactic foams are also available. For example, styrene-butadiene rubber is used to coat hollow particles. These rubber coated particles are incorporated in syntactic foams. This kind of approach can provide a compliant zone around stiff particles and can help in reducing their fracture during compression [71]. Material selection is important in developing such approaches given that the intermediate layer is expected to bond strongly with both hollow particles and the matrix resin. To further enhance such foams, calcium carbonate particles and chopped glass fibers are added along with rubberized hollow particles [72]. Controlling properties of such multiphase composites through optimization of constituent materials is very challenging because of several volume fractions and interfaces involved. Existing studies on these materials are focused on impact and flexural characterization. Compressive properties of such materials can also be interesting due to the presence of several phases.

6.4   Compressive Properties of Nanoscale Reinforced Syntactic Foams

A summary of nanoscale reinforced syntactic foams characterized for compressive properties is presented in Table 6.1. Nanoclay and CNF are the most widely used nanoscale fillers. In general, nanoclay is used in less than 5 vol. % quantity in these composites. Exfoliation of nanoclay becomes difficult at higher content.

The strength and modulus are found to increase almost linearly with nanoclay content in cyanate ester matrix syntactic foams [121]. The compressive strength increased by nearly 250 % and the modulus increased by about 100 % with 4 vol. % nanoclay addition. Use of unexfoliated nanoclay in syntactic foams has shown to improve the compressive failure strain [149]. The compressive energy absorption increased by over 100 % by adding up to 5 vol. % nanoclay in partially dispersed form due to increased failure strain. The presence of nanoclay clusters helped in increasing the failure strain because of sliding of nanoclay platelets with respect to each other. However, presence of clusters results in drawbacks such as reduction in modulus and increase in moisture absorption in composites. The trade-off in properties should be carefully evaluated in such multiscale composites with respect to the requirements of the application. PEEKMOH toughened epoxy matrix syntactic foams containing 40 wt% glass hollow particles showed up to a 33 % increase in compressive strength and 10 % increase in compressive modulus with 5 wt% nanoclay addition [103]. However, the specific strength and modulus were the highest for syntactic foams containing 3 wt% nanoclay. Difficulty in exfoliating nanoclay at higher weight fraction can lead to such observation. Extremely high surface area to volume ratio of nanoclay leads to significant increase in the viscosity of the resin as the nanoclay content is increased and the exfoliation level is improved. The high viscosity may lead to difficulties in syntactic foam fabrication such as increased failure of hollow particle during mixing. The particle failure may lead to high density syntactic foams. Therefore, a careful analysis should be conducted to determine the optimum composition of syntactic foam with the desired set of mechanical properties and density.

The importance of dispersion is further established in a study that used CNTs as reinforcement in phenolic-epoxy matrix syntactic foams [150]. The specimens processed under vacuum were found to have better properties than those processed under normal environmental conditions. CNTs were shown to increase the compressive strength of syntactic foams by 20–60 %. CNTs were incorporated in 0.5, 1, and 1.5 wt% quantities in this study, which resulted in remarkable improvement in compressive properties of syntactic foams. The CNT reinforced syntactic foams showed 40–75 % reduction in compressive properties due to moisture uptake under the conditions of 85 % relative humidity and 71 °C temperature for 30 days. The presence in CNTs helped in better retention of compressive properties after the environmental aging.

Incorporation of CNF has shown to improve the compressive strength of syntactic foams [111, 151]. Reduction in the compressive modulus is observed in this study due to the presence of CNF. It is expected that the cup-stacked structure of CNFs contributes to reduction in the modulus of the composite because the fibers that are not oriented exactly in the direction of loading may easily bend. The curviness along the length of such a long aspect ratio structure limits their ability to increase the compressive stiffness of the composites. Nanofiber pull out is shown as a failure mechanism. Extensive micro and nanoscale deformation is observed in scanning electron micrographs in CNF reinforced epoxy resins and syntactic foams, which helps in improving the energy absorption capability of the material.

In all cases, the hollow particle wall thickness and volume fraction are important parameters. The contribution of nanoparticles is mainly in enhancing the properties of the matrix resin system. The high strain rate compressive deformation characteristics of nanoparticle reinforced syntactic foam systems are not yet available. Strain rate dependence of polymer resins and syntactic foam compressive properties is a topic of research in recent literature [86, 152–154].

6.5   Environmental Effects on Compressive Properties

Environmental conditioning effects on compressive properties of reinforced syntactic foams have been studied for vapor, sea water, and saline water conditions [155]. These conditions are relevant to current marine applications and can provide performance data that can build confidence in reinforced syntactic foams for additional applications. The moisture uptake was higher for the water vapor exposure condition compared to sea and saline water conditions. The moisture that diffuses inside the polymer matrix provides a plasticization effect leading to lower modulus in environmentally exposed syntactic foams compared to their virgin counterparts. The specimens exposed to saline and sea water showed higher compressive strengths, while the specimen under the vaporous condition showed lower strength values compared to the plain syntactic foams containing the same hollow particle type. The decrease in strength of vapor exposed specimens was attributed to the fiber/matrix and hollow particle/matrix interface degradation due to the diffusion of water. Two mechanisms contribute to the interface degradation and reduction in the strength of composites. In the first mechanism, the chemical bonds between the reinforcement and matrix are attacked and weakened by the moisture [156]. In the second mechanism, the difference in the moisture absorption coefficient of fiber and matrix leads to interfacial stresses that cause breaking of bonds and weakening of the interface. In addition, the type of glass used in making hollow particles also plays an important role. The sodalime content, if present in the glass, can leach out and degrade the hollow particle. Hollow particles that are made of only borosilicate glass or other moisture resistant ceramics can improve the environmental performance of syntactic foams.

6.6 Relation Between Density and Compressive Properties

The compressive strength and modulus of various types of syntactic foams are plotted with respect to density in Fig. 6.3 a and b, respectively. Similar to Fig. 5.8, the matrix porosity, material type, and type and content of reinforcement affect density and compressive properties. The graphs show a general trend that with increasing density, the compressive modulus and strength increase. Apart from data sets on a couple of syntactic foam types [118], all other data are covered within a narrow bandwidth the slope of 0.14 and 2.48 MPa/kg/m^3 for strength and modulus, respectively. For a given density value, the compressive modulus and strength vary only over a small range. These graphs are helpful in selecting various constituent materials for obtaining a predetermined set of density and mechanical properties. New methods should be designed to shift these graphs toward lower densities and higher strength values in order to increase the possibility of reducing the structural weight when syntactic foams are used. It is observed in Fig. 6.3 that the maximum compressive strength and modulus of 120 MPa and 2.2 GPa are obtained in reinforced syntactic foams. These properties can be tailored over the entire range up to the maximum possible values. However, the potential of simultaneously tailoring both density and mechanical properties is limited because most of the data fits along a narrow band in this figure.

A comparison of Figs. 6.3 and 5.8 is illustrative of some important aspects related to syntactic foams.[1] The maximum compressive modulus obtained in reinforced syntactic foams is about 2.2 GPa compared to the tensile modulus of about 4.2 GPa. In both cases, CNF reinforced syntactic foams are on the higher end of properties, which indicates the potential for use of nanoscale reinforcement in syntactic foams. The maximum tensile strength is about 35 MPa for any composition but the maximum compressive strength is over 120 MPa. This large difference in tensile and compressive strength clearly reflects in the present applications of syntactic foams, where they are normally used to obtain benefits of compressive properties. One of the reasons for difference in tensile and compressive properties may be that the reinforcing fibers and matrix are the main load bearing elements under tensile loading, while particles can bear load under compression, especially at high volume fractions. The total load bearing cross-section area under tensile loading, which is matrix resin, is much smaller than the actual specimen cross section. In

---

[1] It should be noted that not all published studies have characterized compressive and tensile properties of reinforced syntactic foams and the available data may be obtained on different compositions.

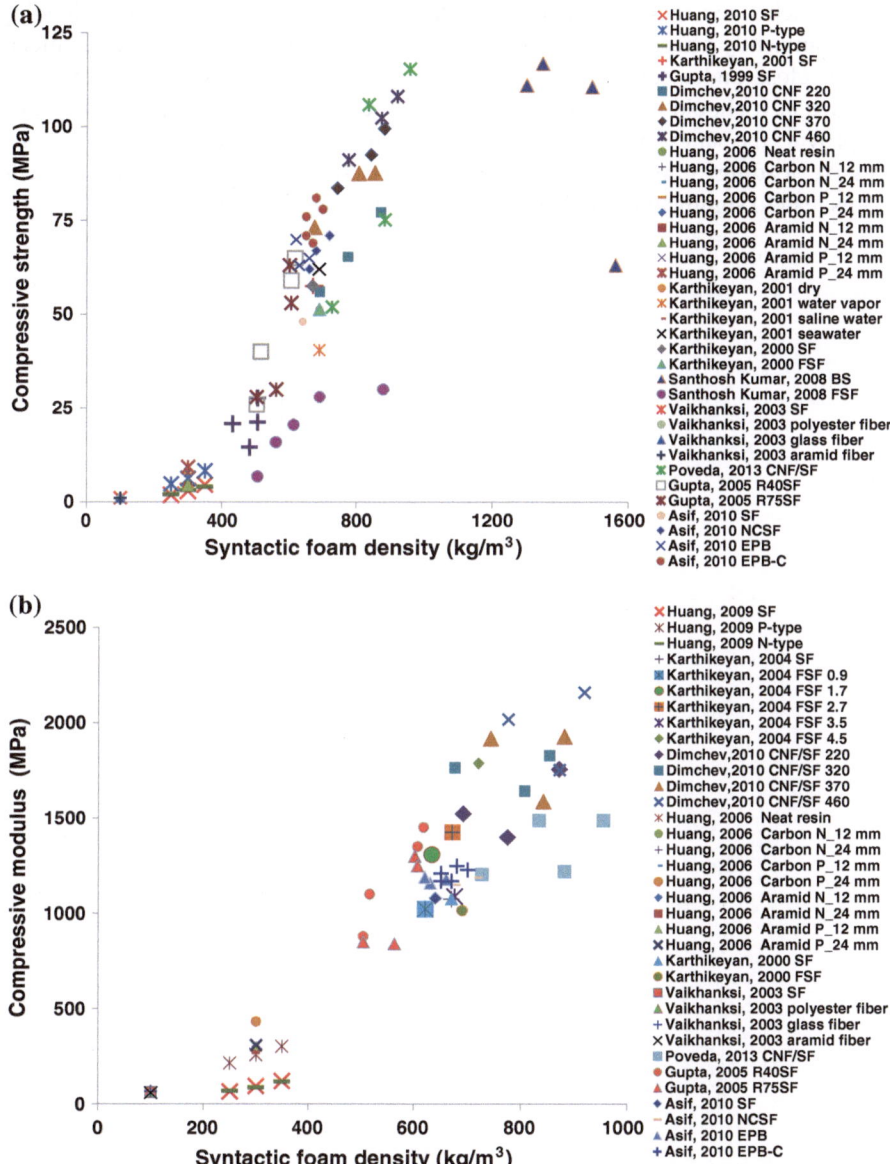

Fig. 6.3   a Compressive strength and **b** modulus of reinforced syntactic foams plotted against the syntactic foam density. Nomenclature: SF plain syntactic foam; FSF fiber reinforced syntactic foam, R40 and R75—rubber particle reinforced syntactic foam; NCSF nanoclay syntactic foam; CNF/SF carbon nanofiber reinforced syntactic foam

addition, the tensile strength of materials has a stronger dependence on defects, such as matrix porosity and particle-matrix debonding, compared to the compressive properties, which may contribute to this observation. Several types of defects such as debonded reinforcement and matrix voids open up under tensile loading and prominently affect the measured tensile strength; on the contrary, these defects tend to close under compression and their effect on compressive strength is not significant. It should also be noted that structure of materials such as CNFs does not provide the same level and mechanism of strengthening under tensile and compressive loading. Large failure strain and high strength of reinforced syntactic foams lead to significantly higher energy absorption under compression.

Chapter 7
Flexural Properties

Abstract The flexural behavior has been studied only for a few reinforced syntactic foams. The short glass fiber reinforced epoxy matrix syntactic foams showed fiber pull out and hollow particle/matrix debonding as the main failure mechanisms under flexural loading conditions. Transition in the failure pattern was observed with the increase in the fiber content. Brittle failure was seen in syntactic foams containing less than 2 vol. % fibers, while foams containing 2–4.5 vol. % fibers showed fiber bending rather than complete failure. Silica particle (8–9 μm diameter) filled syntactic foams showed decreases in flexural strength and modulus with increasing silica content in the range 5–15 wt%. It is found that the syntactic foams with 2 wt% nanoclay show the highest improvement in flexural properties, which include nearly 42 % and 18 % increase in strength and modulus, respectively. The flexural modulus and strength are extracted from the available studies on various reinforced syntactic foams and are plotted with respect to the density. The highest flexural strength and modulus values of any available reinforced syntactic foam are found to be 78 MPa and 3.8 GPa, respectively. Carbon nanofiber reinforced syntactic foams show high compressive and tensile properties but they are not yet tested for flexural properties.

Keywords Syntactic foam • Hollow particle • Nanoscale reinforcement • Glass fiber • Silica fiber • Nanoclay • Flexural modulus • Flexural strength • Density-strength relation • Density-modulus relation

Fiber reinforcement has shown improvement in the tensile and compressive strength and modulus of syntactic foams. These findings indicate possible improvement in the flexural properties as well, because the failure initiation occurs from the tensile side of the syntactic foam specimen subjected to flexural loading when brittle matrix material such as epoxy or vinyl ester is used [137]. However, relatively fewer efforts are found in the published literature related to flexural properties of reinforced syntactic foams, as summarized in Table 7.1. Most of the available data are based on glass hollow particles/epoxy resin syntactic foams reinforced with different types of particles or fibers.

N. Gupta et al., *Reinforced Polymer Matrix Syntactic Foams*, SpringerBriefs in Materials, 53
DOI: 10.1007/978-3-319-01243-8_7, © The Author(s) 2013

Table 7.1 Studies on flexural properties of reinforced syntactic foams

Reference	Matrix resin	Hollow particle material	Reinforcement type
[160]	Epoxy	Glass	Glass fibers
[159]	Epoxy	Glass	Glass fibers
[118]	Benzoxazine	Glass	Silica fibers
[163]	Epoxy	Glass	Nanoclay
[73]	Epoxy	Glass	Nanoclay
[73]	Epoxy	Glass	Crumb rubber
[121]	Cyanate ester	Glass	Nanoclay
[148]	Epoxy	Glass	Crumb rubber with glass fibers and nanoclay
[162]	Vinyl ester	Phenolic	Untreated and silane treated silica micro-particles
[165]	Vinyl ester	Phenolic	Glass, carbon, and basalt fabric

Fig. 7.1 Flexural modulus of fiber reinforced syntactic foams plotted against increasing volume fraction of the fiber content [164]. K20 Glass hollow particles of nominal density 200 kg/m^3

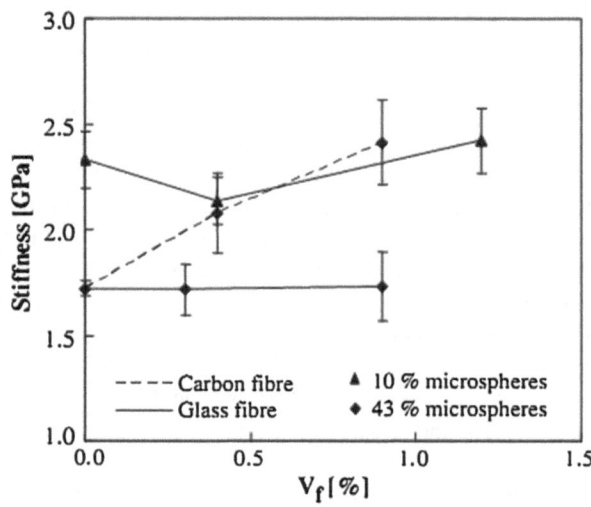

7.1 Fiber Reinforced Syntactic Foam

The flexural and tensile stress–strain graphs of reinforced syntactic foams show similar characteristics (Fig. 4.3), where a linear trend is observed until fracture. The short glass fiber reinforced epoxy matrix syntactic foams showed fiber pull out and hollow particle/matrix debonding as the main failure mechanisms under flexural loading conditions [159, 160]. In general, defects such as matrix voids play a prominent role in defining strength and failure pattern in syntactic foams under flexural loading conditions. Transition in the failure pattern was observed with the increase in the fiber content. Brittle failure was seen in syntactic foams containing less than 2 vol. % fibers, while foams containing 2–4.5 vol. % fibers showed fiber bending rather than complete failure. The flexural modulus of the glass and carbon fiber reinforced syntactic foams is shown in Fig. 7.1, as a function of the fiber volume fraction. The enhancement in mechanical properties due to fibers is sometimes negated by the presence of matrix porosity in syntactic foams [161]. Therefore, a clear understanding of the role and level of mechanical property enhancement caused by fibers is still not present. For example, in an experimental study on syntactic foams with high silica fiber content (9–20 vol. %), the fiber content is reduced and the hollow particle content is increased simultaneously [118]. Therefore, drawing conclusions only for the fiber content is difficult from this work. However, overall density of the composite is affected by both changes and the results showed that the specific strength of the composite remained nearly the same for all compositions.

7.2 Silica Particle Reinforced Syntactic Foam

Silica particle (8–9 μm diameter) filled syntactic foams showed decreases in flexural strength and modulus with increasing silica content in the range 5–15 wt% [162]. Silane treatment of silica particles helped in improving the interfacial bonding and retaining the strength and modulus. Microscopic observation of fracture surface confirmed that the silane treatment improved the interfacial bonding between particles and the vinyl ester matrix.

7.3 Nanoclay Reinforced Syntactic Foam

Effect of nanoclay on flexural behavior of syntactic foams is investigated by adding different amounts of nanoclay in the range of 1–3 wt% [163]. It is found that the syntactic foams with 2 wt% nanoclay show the highest improvement in flexural properties, which include nearly 42 and 18 % increase in strength and modulus, respectively. In a hybrid syntactic foam study containing 0.8 vol. % glass microfibers and 1.6 vol. % nanoclay along with 0, 10, and 20 vol. % crumb rubber particles, flexural properties were measured under four-point bending conditions [148]. The reinforced syntactic foam specimens showed a flexural strength decrease of 24 % and 5 % with the addition of 10 and 20 vol. % crumb rubber, respectively, compared to the plain syntactic foam containing similar composition as shown in Table 7.2.

Table 7.2 Four-point peak bending load data for various specimens tested [148]

	Foam core (N)		Sandwich (N)		Impacted sandwich (N)	
Type	Average	Standard deviation	Average	Standard deviation	Average	Standard deviation
Neat epoxy	3062	74	6270	232	3446	346
RSF	500	80	4842	549	3947	28
RRSF-10	382	15	6236	112	5282	79
RRSF-20	474	4	5731	202	4427	3

RSF Reinforced syntactic foam (nanoclay—1.6 vol.% + microfiber—0.8 vol.%)
RRSF Rubber reinforced syntactic foam (RSF + rubber volume content)

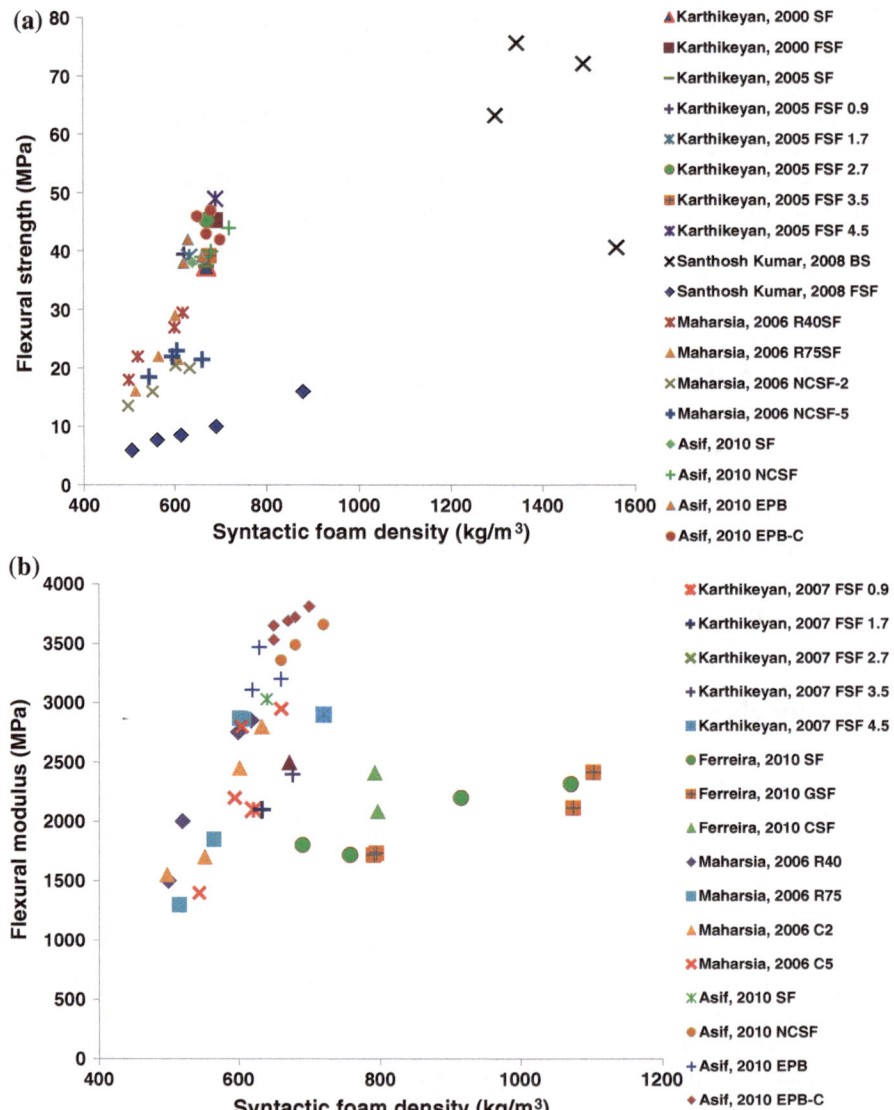

Fig. 7.2 a Flexural strength and **b** modulus of reinforced syntactic foams plotted against the syntactic foam density. Nomenclature : SF plain syntactic foam; FSF fiber reinforced syntactic foam followed by fiber volume fraction; BS Silica fiber reinforced benzoxazine; R40 and R75 rubber particle reinforced syntactic foam; NCSF nanoclay syntactic foam followed by nanoclay volume fraction; EPB syntactic foam with PEEKMOH; EPB-C syntactic foam with PEEKMOH and nanoclay; GSF and CSF glass and carbon fiber reinforced syntactic foam, respectively

7.4 Relations Between Density and Flexural Properties of Syntactic Foams

Flexural strength and modulus values extracted from various studies listed in Table 7.1 are plotted with respect to density in Fig. 7.2. Most reinforced syntactic foam compositions are in the range of 450–800 kg/m^3 density. The highest flexural strength and modulus values of any available reinforced syntactic foam are measured to be 78 MPa and 3.8 GPa, respectively. These values are in between the tensile and compressive property limits on reinforced syntactic foam. CNF reinforced syntactic foams show high compressive and tensile properties but those are not yet tested for flexural properties. The variability in flexural properties is higher than the tensile and compressive properties. Flexural strength is sensitive to surface defects and roughness. The presence of matrix porosity on or near the tensile side of a specimen can result in lower flexural strength and premature failure, which can result in a wider spread in the measured properties. Most of the available data is based on syntactic foams with density in the range of 500–700 kg/m^3. Although the flexural strength and modulus data in Fig. 7.2 are present in a narrow band, the results on one material system may not follow a clear increasing trend over the entire range of densities.

A better understanding of flexural properties requires systematic studies related to changing the fiber and hollow particle contents independently, which has not been done until now. As more compositions of reinforced syntactic foams are characterized for flexural properties and better process control is gained on the manufacturing process, the upper limits for the flexural properties may be raised and the distribution in the data points may be narrowed. Development of reinforced syntactic foams with low matrix porosity is also important in order to attribute the change in mechanical properties to the fiber and hollow particle content.

Chapter 8
Fracture Toughness

Abstract Studies on plain syntactic foams have revealed that the fracture toughness and specific fracture toughness are found to be maximum around 30 vol. % of hollow particles. At low hollow particle volume, fraction stiffening effect and crack bowing failure mechanism was observed whereas at high volume fraction, hollow filler particle-matrix debonding is found to the dominant failure mechanism. Fracture toughness studies on reinforced syntactic foams have been performed only at a constant hollow particle volume fraction of 30 vol. %. A study on phenolic hollow particle filled syntactic foams concluded that fracture toughness increased with increasing fibers content. A maximum increase of 95 % was observed with respect to plain syntactic foam for 10 mm length fiber at 3 wt%. Carbon fibers were found to have a significantly stronger effect on the fracture toughness than glass fibers. PEEKMOH toughened epoxy matrix syntactic foams were found to have up to a 46 % improvement in fracture toughness with the addition of 5 wt% nanoclay. An increase of 37 % in fracture toughness is observed for the addition of 1.5 vol. % of carbon nanofibers in comparison to plain syntactic foams. It was also observed that microscale reinforcement (short carbon fibers) was more effective than nanoscale reinforcement (nanoclay), at similar weight fractions.

Keywords Syntactic foam • Hollow particle • Nanoscale reinforcement • Carbon fiber • Carbon nanotube • Nanoclay • Glass fiber • Fracture toughness • Density-fracture toughness relation

To characterize syntactic foams for fracture toughness, the single edge notched specimen geometry containing a sharp crack is usually tested under three-point bend conditions. Fracture toughness studies on reinforced and plain syntactic foams are available in the published literature. The characterization of plain syntactic foams for fracture toughness, under varying volume fractions and types of hollow particles have revealed that the toughness and specific fracture toughness are found to be maximum around 30 vol. % of hollow particles [166, 167].

N. Gupta et al., *Reinforced Polymer Matrix Syntactic Foams*, SpringerBriefs in Materials,   59
DOI: 10.1007/978-3-319-01243-8_8, © The Author(s) 2013

Table 8.1 Fracture toughness studies on reinforced syntactic foams

Reference	Matrix resin	Hollow particle material	Reinforcement type
[103]	Epoxy + PEEKMOH (1–5 wt%)	Glass ($\Phi_{mb} =$ 0.35–0.4 wt%)	Nanoclay (1–5 wt%)
[115]	Epoxy	Phenolic ($\Phi_{mb} = 0.3$)	Short carbon fiber- 10 mm (0–3 wt%), nanoclay (0–2 wt%)
[120]	Epoxy	Phenolic ($\Phi_{mb} = 0.3$)	Short carbon fiber-3.11, 4.5 and 10.05 mm (0–3 wt%)
[164]	Epoxy	Glass ($\Phi_{mb} = 0$–0.5)	E-glass fiber-3 mm and carbon fiber-3 mm
[168]	Phenolic	Carbon ($\Phi_{mb} = 0.28$)	Carbon nanofiber-(0.5–2 %)

The studies suggest that a change in the failure mechanism of the syntactic foams occurs around hollow particle volume fraction of 0.3. At low Φ_{mb}, a stiffening effect and crack bowing mechanism can be observed, while hollow filler particle-matrix debonding is found to be the dominant failure mechanism at high Φ_{mb} [167].

The studies characterizing the effect of reinforcement on the fracture toughness of syntactic foams have usually been performed at a constant Φ_{mb}. Reinforcement is expected to increase the fracture toughness of syntactic foams by one or more of the following mechanisms:

• Crack bridging by long aspect ratio reinforcements
• Energy dissipation in the fracture of an additional phase and
• Increase in the crack path length due to the presence of reinforcement.

The existing literature on the fracture toughness of plain and reinforced syntactic foams is given in Table 8.1. The chapter discusses the effect of nanoscale reinforcements, followed by microscale reinforcements, and concludes with a comparison between the nano and microscale reinforcements on the fracture toughness of syntactic foams.

8.1 Nanoscale Reinforced Syntactic Foams

The nanoscale reinforcements used include nanoclay and carbon nanofiber. The effect of nanoclay (1–5 wt%) was studied on glass hollow particle/PEEKMOH toughened epoxy matrix syntactic foams. The fracture toughness of epoxy matrix syntactic foams is found to increase by 26 % through the addition of 5 wt% nanoclay [103]. PEEKMOH toughened epoxy matrix syntactic foams were found to have up to a 46 % improvement in fracture toughness at 5 wt% nanoclay addition. The improvement in the fracture toughness due to the PEEKMOH, is attributed to the phase separation of the PEEKMOH domains in the epoxy matrix [103].

The effect of CNFs was studied on carbon hollow particle/phenolic resin syntactic foam. An increase of ~37 % in fracture toughness is observed for the addition of 1.5 vol. % of CNFs, in comparison to the plain syntactic foam. Both fiber pullout and crack bowing mechanisms were observed as the failure mechanisms in reinforced and plain syntactic foams, respectively.

8.2 Fiber Reinforced Syntactic Foams

The effect of fiber length on the fracture toughness of phenolic hollow particle filled syntactic foams has been studied [120]. To perform this analysis, the volume fraction of hollow particles was kept at a constant 30 %. For all the fiber lengths tested (3, 4.5 and 10 mm), the fracture toughness increased with increasing fibers content and a maximum increase of 95 % was observed for 10 mm length fiber at 3 wt%, with respect to plain syntactic foam. The effect of various types of fiber materials such as glass and carbon fibers on the fracture toughness of syntactic foams has been studied [164]. Carbon fibers were found to have a significantly stronger effect on the fracture toughness than glass fibers. Addition of 0.9 vol. % of carbon fibers was found to improve the fracture toughness by 35 % compared to the plain syntactic foams. Fiber pull-out, particle matrix debonding and particle failure were observed as the failure mechanisms. The addition of fiber reinforcements created step-like structures during the failure process, which helped in enhancing the fracture toughness due to the high surface area of such structures.

8.3 Comparison of Nano and Microscale Toughening

The effect of nanoscale (nanoclay) and microscale reinforcements (short carbon fiber) on the fracture toughness of phenolic hollow particle filled epoxy matrix syntactic foams has been studied [115]. The study was performed with keeping the volume fraction of hollow particles constant at 30 % and varying the nanoclay content from 0–2 wt% and the carbon fiber content from 1–3 wt%. A 42 % increase in the fracture toughness was observed with the addition of 1 wt% nanoclay in comparison to plain syntactic foam. With further increase in the nanoclay content, a decrease in the fracture toughness was observed, owing to the agglomeration of the nanoclay layers. The energy release rate was found to have a maximum increase of 104 % at 1 wt% of the nanoclay reinforcement, compared to plain syntactic foam. The SCF reinforced syntactic foams showed an increasing fracture toughness and energy release rate with an increasing weight fraction of carbon fibers. A maximum increase of about 109 % in the fracture toughness is observed at 3 wt% of the fiber addition. The study also observed microscale reinforcement (SCF) was more effective than nanoscale reinforcement (nanoclay), under similar weight fractions.

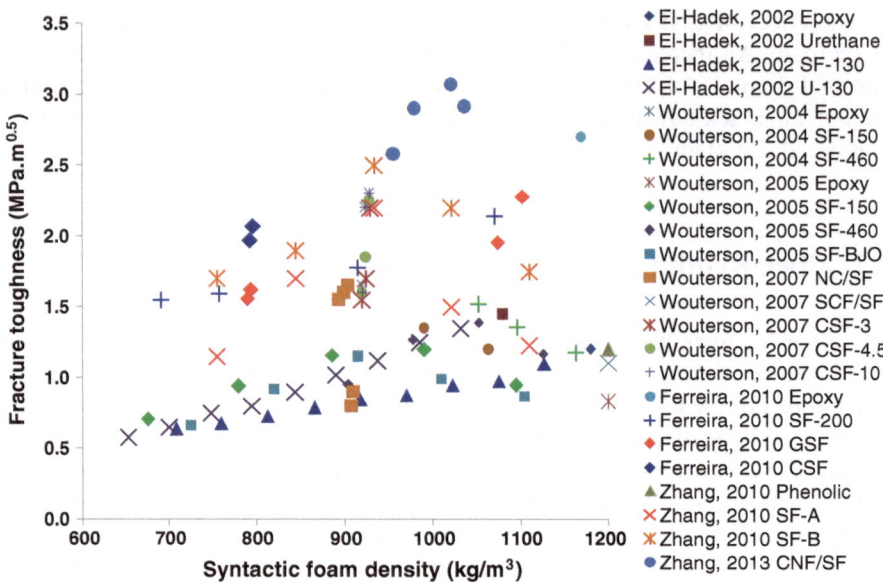

Fig. 8.1 Fracture toughness for plain and reinforced syntactic foams plotted as a function of the syntactic foam density

Irrespective of the addition of different types of reinforcement to augment the fracture toughness, the primary consideration for syntactic foams is the weight saving potential. To fully utilize this important aspect of the composite, the fracture toughness values of all available plain and reinforced syntactic foams are plotted as a function of the syntactic foams density in Fig. 8.1. In most syntactic foam systems, the fracture toughness is found to increase with the syntactic foam density. Figure 8.1 shows that several syntactic foam compositions can be found having the same fracture toughness but different densities. The fracture toughness varies between 0.6 and 3 MPa.m$^{0.5}$.

Chapter 9
Dynamic Mechanical Properties

Abstract Studying the influence of temperature and loading frequency on the behavior of syntactic foams is important because of its diverse set of applications. Dynamic mechanical analysis (DMA) is a widely used technique for measuring viscoelastic properties of materials over a range of temperatures and loading frequencies. The storage modulus and loss modulus determined in a DMA experiment measure the capacity of a material to store and dissipate energy, respectively. In general, the storage modulus of syntactic foams decreases with increasing temperature. This response was consistent between plain and reinforced syntactic foams. Study of storage modulus of vinyl ester/glass hollow particle syntactic foams at three different temperatures concluded that the neat resin has higher storage modulus than the syntactic foams below glass transition temperature (T_g) but this trend is reversed above T_g. Also, the room temperature (30 °C) storage modulus of syntactic foams increases with the increase in the wall thickness of hollow particles. The addition of nanoclay increased the storage modulus of epoxy matrix syntactic foams. The effect was attributed to the toughening of matrix resin by nanoclay particles. However, increased stiffness of nanoclay reinforced syntactic foams resulted in decreased loss modulus. Cyanate ester matrix syntactic foam with 4 vol. % of nanoclay showed higher storage modulus than the plain syntactic foams, owing to the restricted movement of the polymer chains which can be attributed to the good interaction between the nanoclay and the matrix resin.

Keywords Syntactic foam • Hollow particle • Nanoscale reinforcement • Glass fiber • Carbon fiber • Carbon nanotube • Nanoclay • Storage modulus • Loss modulus • Damping parameter • Glass transition temperature • Loading frequency

Dynamic mechanical analysis (DMA) can provide viscoelastic properties, namely storage modulus, loss modulus, and the damping parameter (Tan δ) of materials. The dynamic mechanical properties can be studied by a dynamic mechanical analyzer at different temperatures and loading frequencies. DMA is a very versatile technique

N. Gupta et al., *Reinforced Polymer Matrix Syntactic Foams*, SpringerBriefs in Materials, DOI: 10.1007/978-3-319-01243-8_9, © The Author(s) 2013

that can provide a number of material parameters in a single set of experiments. The experiments can be conducted in several configurations, such as tension, compression, three- or four-point bending, and vibration of single and dual cantilever.

The available literature on the DMA of reinforced syntactic foams is summarized in Table 9.1. The existing literature on reinforced syntactic foam has mostly studied the variation of the viscoelastic properties with respect to temperature at a constant frequency of 1 Hz. Three-point bending is the most common specimen loading configuration during the test. In a DMA experiment, the material is subjected to a sinusoidal loading and the response of the material is recorded. The in-phase and the out-of-phase component of the stress are utilized to evaluate the storage and the loss modulus [169]. The ratio of the loss modulus to the storage modulus is evaluated as the damping parameter. The storage modulus gives a measure of the energy stored in the material, while the loss modulus gives the energy dissipated in each loading cycle. The DMA testing is also capable of providing measurement of glass transition temperature (T_g), which is estimated as the temperature at which the loss modulus value shows a peak.

The storage modulus-temperature profiles of plain and reinforced syntactic foams are similar. In general, with increasing temperature, the storage modulus of syntactic foams decreases. A typical storage modulus-temperature profile is shown in Fig. 9.1a. The curve can be divided into three regions. The region I is characterized

Table 9.1 Dynamic mechanical properties of reinforced syntactic foams

References	Matrix resin	Hollow particle	Reinforcement	Test conditions[a]
[172]	Epoxy	Glass	Glass fibers	Tensile mode, ω and a = unspecified, $T = 20$–100 °C
[170]	Epoxy	Glass	Glass and carbon fibers	Three-point bending mode, ω and a = unspecified, $T = 20$–100 °C
[121]	Cyanate ester	Glass	Nanoclay	Bending mode, $\omega = 1$ Hz, $T = 75$–325 °C
[103]	Epoxy resin, PEEKMOH toughened epoxy	Glass	Nanoclay	Three-point bending mode, $\omega = 1$ Hz, $T =$ R.T.–300 °C
[120]	Epoxy	Phenolic	Carbon fibers	Three-point bending mode, $\omega = 1$ Hz, a = unspecified and $T =$ R.T.–110 °C
[150]	Phenolic-epoxy	Glass	Carbon nanotubes	Compression mode, $\omega = 1$ Hz, $a = 0.02$ % and $T = 50$–180 °C

[a] T temperature, R. T. room temperature, ω frequency, a amplitude

by gradual decrease in the storage modulus, region II represents the glass transition region and is characterized by drastic reduction in the storage modulus, and region III is the flow region where the storage modulus remains nearly constant at a low value. Figure 9.1b shows the response of a variety of reinforced syntactic foams, where the curves for all material types show a similar general trend. The storage modulus values for a variety of vinyl ester/glass hollow particle syntactic foams at three different temperatures are provided in Table 9.2 for illustrative purposes. The neat resin has higher storage modulus than the syntactic foams below T_g but this trend is reversed above T_g. The specific storage modulus for syntactic foams may be higher than the neat resin as per the results shown in the table. The room temperature (30 °C) storage modulus of syntactic foams increases with the increase in the wall thickness of the microballoons. The room temperature storage modulus and

Fig. 9.1 a Variation of the storage modulus of neat vinyl ester resin showing the three regions [171] and **b** variation of the storage modulus of carbon fiber reinforced syntactic foams with respect to temperature [120]

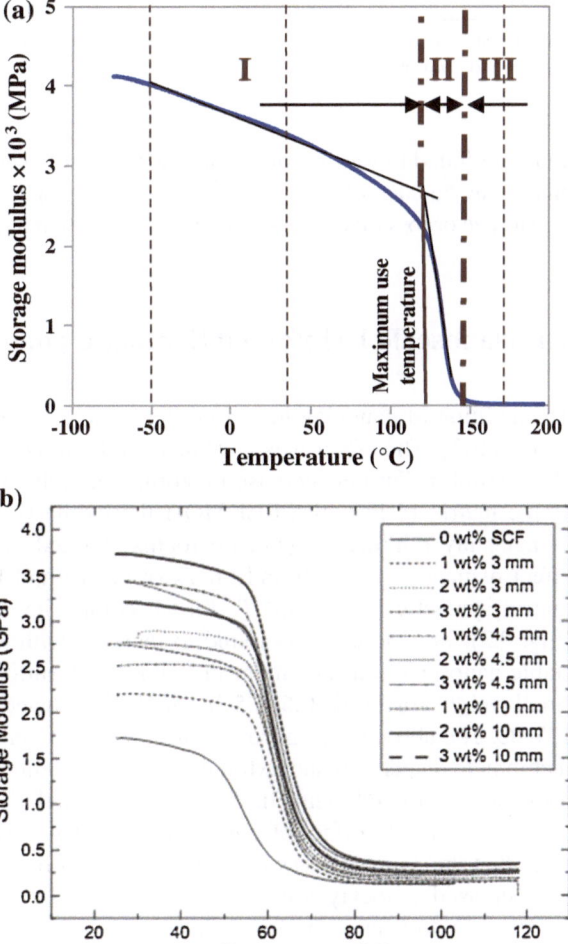

Table 9.2 Comparison of storage modulus values for vinyl ester/glass hollow particle syntactic foams at various temperatures

Specimen type[a]	E' at −50 °C (MPa)	E' at 30 °C (MPa)	E' at 175 °C (MPa)
Neat VE	4,041 ± 44	3,457 ± 42	11 ± 1
VE220-30	2,916 ± 76	2,548 ± 66	50 ± 2
VE220-40	2,841 ± 54	2,467 ± 34	72 ± 0
VE220-50	3,015 ± 42	2,640 ± 44	124 ± 3
VE220-60	2,186 ± 65	1,998 ± 37	178 ± 2
VE320-30	3,554 ± 45	3,065 ± 53	46 ± 2
VE320-40	3,369 ± 49	2,949 ± 55	71 ± 2
VE320-50	3,139 ± 73	2,810 ± 98	126 ± 3
VE320-60	2,935 ± 55	2,628 ± 63	213 ± 7
VE460-30	4,082 ± 24	3,523 ± 36	51 ± 1
VE460-40	3,755 ± 143	3,311 ± 120	85 ± 9
VE460-50	3,557 ± 206	3,264 ± 186	137 ± 7
VE460-60	3,712 ± 94	3,410 ± 36	248 ± 25

[a]Specimen nomenclature: VE denotes vinyl ester matrix, followed by the glass hollow particle density and volume %

Tan δ are found to vary linearly with the density of syntactic foams. The reinforcement phase may be added to syntactic foams with any specific objective such as increasing the storage or loss modulus based on the application requirements.

9.1   Nanoscale Reinforced Syntactic Foams

The addition of nanoclay increased the storage modulus of epoxy matrix syntactic foams [103]. The effect was attributed to the toughening of matrix resin by nanoclay particles. Further increase in storage modulus was observed with toughening of the matrix resin with PEEKMOH thermoplastic. However, increased stiffness of nanoclay reinforced syntactic foams resulted in decreased loss modulus. No effect of nanoclay was found on T_g of epoxy and PEEKMOH toughened epoxy matrix syntactic foams and T_g for all composites was recorded around 200 °C (Fig. 9.2). In general, T_g is found to increase with addition of a high temperature stable nanoscale reinforcement in polymers. Reinforcement of epoxy matrix with small weight fractions (0.5–1.5 %) of CNTs showed an increase in T_g from 112 to 137 °C. The absence of improvement in T_g due to nanoclay incorporation indicates incomplete dispersion and exfoliation. The maximum use temperature (T_{max}) of all nanoclay reinforced syntactic foams was around 175 °C compared to 160 °C for the neat resin. No effect of nanoclay content was observed on T_{max}, which indicates that the nanoclay was not completely exfoliated and the potential gain due to the increased nanoclay content was neutralized by the presence of clusters.

The effect of nanoclay on cyanate ester matrix syntactic foams in the temperature range of 75–325 °C has been studied [121]. Syntactic foam with 4 vol. % of

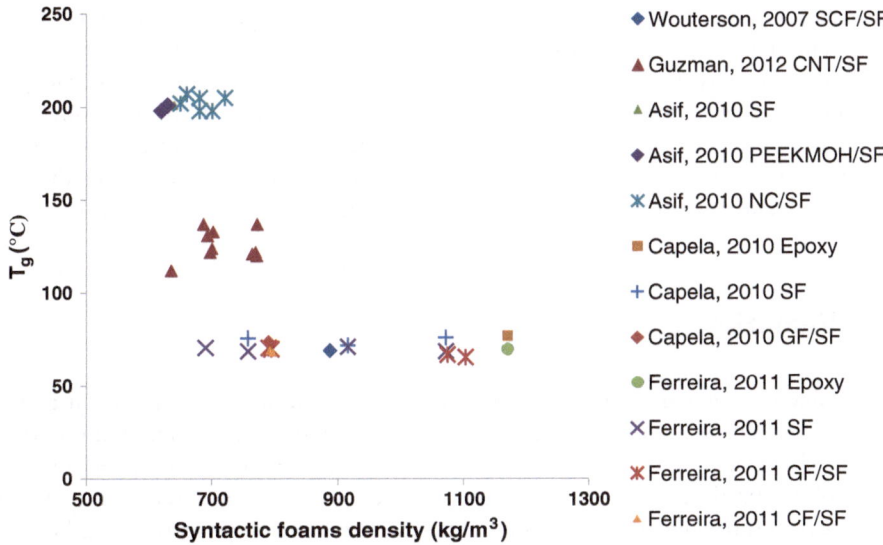

Fig. 9.2 The glass transition temperature for the various types of syntactic foams presents as a function of the syntactic foam density

nanoclay showed higher storage modulus than the plain syntactic foam, owing to the restricted movement of the polymer chains which can be attributed to the good interaction between the nanoclay and the matrix resin. The glass transition temperature decreased to 170 from 275 °C, as the volume fraction of the nanoclay increased to 4 vol. %. The intercalated morphology of the nanoclay restricts the resin mobility, thereby decreasing the damping parameter as the volume fraction increases. The effect of another type of nanoscale reinforcement, namely carbon nanotubes, on the viscoelastic properties of syntactic foam have also been studied [150]. The syntactic foam containing 0.5 wt% of functionalized multiwalled carbon nanotube reinforcement showed a 22 % increase in the glass transition temperature.

9.2 Microscale Reinforced Syntactic Foams

The effects of glass and carbon fibers on the viscoelastic properties of syntactic foams have been studied using three-point bend configuration of the dynamic mechanical analyzer [170]. In Region I, the storage modulus of both the fiber reinforced syntactic foams was higher than the neat resin for 2 and 13 wt% of hollow particles. The addition of fibers and hollow particles did not significantly alter the glass transition and the maximum use temperatures from the neat resin values. At room temperature (25 °C), the syntactic foams containing 13 wt% of hollow particles and reinforced with glass fibers showed 19–38 % of higher storage modulus than carbon fiber reinforcements. The damping parameter decreased for both the fiber reinforced syntactic foams in comparison to the neat resin, with 2 wt% carbon fiber reinforced syntactic foams having the lowest value. The effect of fiber length on the viscoelastic properties of syntactic foams showed that with increasing volume fraction of the fiber content, the storage modulus increased in comparison to the plain syntactic foam [120]. The variation of the fiber length and the quantity did not significantly affect the glass transition temperature of the syntactic foam.

9.3 Remarks

The value of syntactic foams is in their ability to be tailored for the given application. The diverse set of applications may require syntactic foams with a wide range of viscoelastic properties. The availability of numerous parameters including hollow particle wall thickness and volume fraction, reinforcement material size, aspect ratio, and volume fraction and the matrix resin properties can help in designing the syntactic foams that are suitable for the application. However, this large set of available variables may also require considerable work in narrowing down the parameters to obtain the given set of properties. It appears from the results that nanoscale fillers have a strong effect on the glass transition temperature of syntactic foams. The effect of microscale fillers is not prominent. High surface area to volume ratio of nanomaterials that can promote a larger number of interfacial bonds than the microscale fillers may be responsible for this observation. The observation about T_g indicates that the thermal properties of syntactic foams can also be modulated by using an additional phase in the composite.

Chapter 10
Summary and Future Challenges

Abstract In the previous chapters, the existing literature on the mechanical properties of reinforced syntactic foams has been reviewed. This chapter summarizes the effects of the reinforcements on the mechanical properties of syntactic foams and identifies the critical areas where a lack of literature is observed relating to reinforced syntactic foams. The data for various mechanical properties such as compressive, tensile, and the flexural strength and modulus have been plotted as a function of the composites density. This helps in identifying the composite compositions that can be helpful in weight saving applications. The nanoscale reinforcements help in obtaining the highest mechanical properties without increasing the density of syntactic foams. These results show that the nanoscale reinforcements can push the boundaries of mechanical properties of syntactic foams. It is identified that understanding the effect of moisture, weathering, and ultraviolet light radiation on the composite is not well developed. The lack of theoretical models for reinforced syntactic foams, considering the interaction between the nanoscale reinforcement and the hollow particles, is also noted in the literature and should be the focus of future research efforts.

Keywords Syntactic foam • Hollow particle • Density • Porosity • Nanoscale reinforcement • Carbon nanofiber • Carbon nanotube • Nanoclay • Compressive modulus • Strength • Modulus

Various types of micro and nano-sized reinforcements have been used in syntactic foams to improve their mechanical properties. A comparative study of mechanical properties of reinforced syntactic foams is conducted in the present review. In all cases, the experimentally measured mechanical properties are plotted against density of the composite. The tensile and compressive strength and modulus show a nearly linear trend, within a narrow band, plotted with respect to density without any regard to the type of materials used in fabricating the syntactic foam. The flexural strength and modulus show a larger range of scatter in the mechanical properties plotted with respect to density. In recent years, considerable effort

N. Gupta et al., *Reinforced Polymer Matrix Syntactic Foams*, SpringerBriefs in Materials,   69
DOI: 10.1007/978-3-319-01243-8_10, © The Author(s) 2013

has been invested in developing processing methods that can reduce the entrapped matrix porosity or improve particle-matrix interfacial bonding with the expectation of improving the mechanical properties of such foams. However, the available trends show that the potential benefits of such methods are limited when the density of the composites are factored in. Nanoscale reinforcement is successful in pushing the upper limits of mechanical properties obtained in syntactic foams. The highest levels of tensile and compressive properties are observed in nanoscale reinforced syntactic foams, although such composites also have high density. Similar observations are found in flexural properties of reinforced syntactic foams. Further improvements in strength and modulus can be obtained by optimizing the nanoscale reinforcement type and content. Methods to obtain complete dispersion of nanoscale fillers in matrix resins will help in obtaining uniform properties in syntactic foams.

Orienting the reinforcements in a desired direction to obtain the required mechanical properties in reinforced syntactic foams is a challenge for the current fabrication methods. Several upcoming applications of syntactic foams, especially as the core in sandwich structures, will benefit from oriented reinforcements because such structures are often designed for compressive loading conditions. Interest in the impact and high strain rate compressive properties of plain and reinforced syntactic foams is also increasing in recent years. Compressive properties and failure mechanisms of plain syntactic foams are strain rate dependent. However, the state of knowledge for reinforced syntactic foams under such loading conditions is in a nascent stage at this point and only a few isolated attempts are discovered [94, 173, 174]. Systematic experimental studies and availability of modeling tools for understanding the strain rate dependence of mechanical properties of reinforced syntactic foams are required by their current and future applications.

An area of immediate need is the development of theoretical models for reinforced syntactic foams. Modeling the mechanical behavior of microfiber reinforced syntactic foams is very challenging because a unit cell-based geometry for this kind of structure lacks general symmetry. The theoretical studies should also take into account the fiber-particle interaction depending on the type of reinforcement used. The interaction effects can be significant in composites and can lead to under- or over-estimation of properties depending on factors such as relative stiffness, size, and volume fractions of constituents.

Theoretical models are now available to estimate the tensile and flexural modulus of plain syntactic foams. These models are validated with experimental results on a wide range of syntactic foams and can be applied to nanofiber and nanoparticle reinforced composites through innovative methods as demonstrated in this study. These models take the experimentally measured properties of nanoparticle reinforced resins as input for the matrix properties and provide estimates for the overall composite properties. Such schemes are successful in predicting the properties of CNF and nanoclay reinforced syntactic foams. However, molecular level modeling or simulation methods that can predict the properties of nanoparticle reinforced resins and further extend the modeling capabilities need to be interfaced with microscale homogenization techniques. Such multiscale methods will lead to theoretical predictions at all levels and eliminate the need of obtaining experimental measurements of nanoparticle reinforced resins.

Low density of syntactic foams has resulted in several marine and aerospace applications, which can benefit from detailed studies on environmental degradation of reinforced syntactic foams. Only a few reinforced syntactic foam systems have been exposed to environmental degradation conditions and understanding of the degradation mechanism, residual mechanical properties, and failure mechanisms is not yet present. Practical applications will immensely benefit from the presence of data related to these aspects. Long-term environmental studies can include characterizing the effects of moisture, temperature, and ultra-violet radiation on the degradation mechanism of syntactic foams.

Advancements related to development of new nanomaterials, nanomaterial characterization techniques, imaging technologies, dispersing nanoparticles in polymers, and creating tailored interfaces are helping numerous fields. Reinforced syntactic foams will also directly benefit from these research fields. The ability to tailor syntactic foams over a wide range of density and mechanical properties will be enhanced by these advancements.

References

1. Hodge AJ, Kaul RK and McMahon WM (2000) Sandwich composite, syntactic foam core based application for space structures. In Loud S et al (ed) Proceedings of the 45th International SAMPE Symposium, 21–25 May 2000. SAMPE Publishing, Long Beach, p. 2293–2304
2. Rohatgi PK, Gupta N, Schultz BF, Luong DD (2011) The synthesis, compressive properties, and applications of metal matrix syntactic foams. JOM J Miner Met Mater Soc 63(2):30–36
3. Shutov F (1986) Syntactic polymer foams, in Advanced Polymer Science. Springer, New York, pp 63–123
4. Rutz BH, Berg JC (2010) A review of the feasibility of lightening structural polymeric composites with voids without compromising mechanical properties. Adv Colloid Interface Sci 160(1–2):56–75
5. Gupta N, Ye R, Porfiri M (2010) Comparison of tensile and compressive characteristics of vinyl ester/glass microballoon syntactic foams. Compos B Eng 41(3):236–245
6. Grosjean F, Bouchonneau N, Choqueuse D, Sauvant-Moynot V (2009) Comprehensive analyses of syntactic foam behaviour in deepwater environment. J Mater Sci 44(6):1462–1468
7. Hobaica EC, Cook SD (1968) The characteristics of syntactic foams used for buoyancy. J Cell Plast 4(4):143–148
8. National Oceanic and Atmospheric Administration, http://oceanexplorer.noaa.gov/technology/subs/alvin/alvin.html
9. Wikipedia.com, http://en.wikipedia.org/wiki/File:Zumwalt_Deckplate_Transit.jpg
10. Cochran JK (1998) Ceramic hollow spheres and their applications. Curr Opin Solid State Mater Sci 3(5):474–479
11. Shabde V, Hoo K, Gladysz GM (2006) Experimental determination of the thermal conductivity of three-phase syntactic foams. J Mater Sci 41(13):4061–4073
12. Huang Y-J, Wang C-H, Huang Y-L, Guo G, Nutt SR (2010) Enhancing specific strength and stiffness of phenolic microsphere syntactic foams through carbon fiber reinforcement. Polym Compos 31(2):256–262
13. Rohatgi PK, Matsunaga T, Gupta N (2009) Compressive and ultrasonic properties of polyester/fly ash composites. J Mater Sci 44(6):1485–1493
14. Kulkarni S, Kishore (2003) Effect of filler–fiber interactions on compressive strength of fly ash and short-fiber epoxy composites. J Appl Polym Sci 87(5):836–841
15. Fomenko EV, Anshits NN, Pankova MV, Solovyov LA, Anshits AG (2011) Fly ash cenospheres: composition, morphology, structure, and helium permeability. World of Coal Ash (WOCA) Conference, 9–12 May 2011

16. Matsunaga T, Kim JK, Hardcastle S, Rohatgi PK (2002) Crystallinity and selected properties of fly ash particles. Mater Sci Eng, A 325(1–2):333–343

17. Carlisle KB, Lewis M, Chawla KK, Koopman M, Gladysz GM (2007) Finite element modeling of the uniaxial compression behavior of carbon microballoons. Acta Mater 55(7):2301–2318

18. Koopman M, Gouadec G, Carlisle KB, Chawla KK, Gladysz GM (2004) Compression testing of hollow microspheres (microballoons) to obtain mechanical properties. Scripta Mater 50(5):593–596

19. Wang DJ, Zhang YH, Dong AG, Tang Y, Wang YJ, Xia JC, Ren N (2003) Conversion of fly ash cenosphere to hollow microspheres with zeolite/mullite composite shells. Adv Funct Mater 13(7):563–567

20. Gupta N, Brar BS, Woldesenbet E (2001) Effect of filler addition on the compressive and impact properties of glass fibre reinforced epoxy. Bull Mar Sci 24(2):219–223

21. Sudarshan, Surappa MK (2008) Synthesis of fly ash particle reinforced A356 Al composites and their characterization. Mater Sci Eng, A, 2008 480(1–2):117–124

22. Kulkarni S, Anuradha D, Murthy C, Kishore (2002) Analysis of filler-fibre interaction in fly ash filled short fibre-epoxy composites using ultrasonic NDE. Bull Mar Sci 25(2):137–140

23. Gu J, Wu G, Zhao X (2008) Damping properties of fly ash/epoxy composites. J Univ Sci Technol Beijing, Mineral, Metall, Mater 15(4):509–513

24. Sun L-H, Ounaies Z, Gao X-L, Whalen CA, Yang Z-G (2011) Preparation, characterization, and modeling of carbon nanofiber/epoxy nanocomposites. J Nanomater 2011:1–8

25. Chaos-Morán R, Salazar A, Ureña A (2011) Mechanical analysis of carbon nanofiber/ epoxy resin composites. Polym Compos 32(10):1640–1651

26. Karippal JJ, Narasimha Murthy HN, Rai KS, Sreejith M, Krishna M (2011) Study of mechanical properties of epoxy/glass/nanoclay hybrid composites. J Compos Mater 45(18):1893–1899

27. Lau K-T, Gu C, Hui D (2006) A critical review on nanotube and nanotube/nanoclay related polymer composite materials. Compos B Eng 37(6):425–436

28. Patton RD, Pittman JCU, Wang L, Hill JR (1999) Vapor grown carbon fiber composites with epoxy and poly(phenylene sulfide) matrices. Compos A Appl Sci Manuf 30(9):1081–1091

29. Tibbetts GG, Lake ML, Strong KL, Rice BP (2007) A review of the fabrication and properties of vapor-grown carbon nanofiber/polymer composites. Compos Sci Technol 67(7–8):1709–1718

30. Gupta N, Woldesenbet E, Mensah P (2004) Compression properties of syntactic foams: effect of cenosphere radius ratio and specimen aspect ratio. Compos A Appl Sci Manuf 35(1):103–111

31. Porfiri M, Gupta N (2009) Effect of volume fraction and wall thickness on the elastic properties of hollow particle filled composites. Compos B Eng 40(2):166–173

32. Zainuddin S, Hosur MV, Zhou Y, Narteh AT, Kumar A, Jeelani S (2010) Experimental and numerical investigations on flexural and thermal properties of nanoclay-epoxy nanocomposites. Mater Sci Eng A 527(29–30):7920–7926

33. Pinnavia TJ, Beall GW (2000) Polymer-clay nanocomposites. Wiley Series in Polymer Science. John Wiley, New York, 350 p

34. Gupta N, Lin T, Shapiro M (2007) Clay-epoxy nanocomposites: Processing and properties. JOM J Miner, Met Mater Soc 59(3):61–65

35. Chan M-L, Lau K-T, Wong T-T, Ho M-P, Hui D (2011) Mechanism of reinforcement in a nanoclay/polymer composite. Compos B Eng 42(6):1708–1712

36. Choudalakis G, Gotsis AD (2009) Permeability of polymer/clay nanocomposites: A review. Eur Polymer J 45(4):967–984

37. Khan SU, Munir A, Hussain R, Kim J-K (2010) Fatigue damage behaviors of carbon fiber-reinforced epoxy composites containing nanoclay. Compos Sci Technol 70(14):2077–2085

38. Liu J, Boo WJ, Clearfield A, Sue HJ (2006) Intercalation and exfoliation: A review on morphology of polymer nanocomposites reinforced by inorganic layer structures. Mater Manuf Processes 21(2):143–151

39. Maniar KK (2004) Polymeric Nanocomposites: A Review. Polymer–Plastics Technology and. Engineering 43(2):427–443
40. Paul DR, Robeson LM (2008) Polymer nanotechnology: Nanocomposites. Polymer 49(15):3187–3204
41. Pavlidou S, Papaspyrides CD (2008) A review on polymer-layered silicate nanocomposites. Prog Polym Sci 33(12):1119–1198
42. Li H, Zhong J, Meng J, Xian GJ (2013) The reinforcement efficiency of carbon nanotubes/ shape memory polymer nanocomposites. Compos B Eng 44(1):508–516
43. Feng Q-P, Shen X-J, Yang J-P, Fu S-Y, Mai Y-W, Friedrich K (2011) Synthesis of epoxy composites with high carbon nanotube loading and effects of tubular and wavy morphology on composite strength and modulus. Polymer 52(26):6037–6045
44. Baughman RH, Zakhidov AA, de Heer WA (2002) Carbon nanotubes–The route toward applications. Science 297(5582):787–792
45. Breuer O, Sundararaj U (2004) Big returns from small fibers: A review of polymer/carbon nanotube composites. Polym Compos 25(6):630–645
46. Iijima S (2002) Carbon nanotubes: past, present, and future. Phys B Phys Condens Matter 323(1–4):1–5
47. Maruyama B, Alam K (2002) Carbon nanotubes and nanofibers in composite materials. SAMPE J 38(3):59–70
48. Sharma SP, Lakkad SC (2011) Effect of CNTs growth on carbon fibers on the tensile strength of CNTs grown carbon fiber-reinforced polymer matrix composites. Compos A Appl Sci Manuf 42(1):8–15
49. Sinnott SB, Andrews R (2001) Carbon nanotubes: synthesis, properties, and applications. Crit Rev Solid State Mater Sci 26(3):145–249
50. Sun Y-P, Fu K, Lin Y, Huang W (2002) Functionalized carbon nanotubes: Properties and applications. Acc Chem Res 35(12):1096–1104
51. Allaoui A, Bai S, Cheng HM, Bai JB (2002) Mechanical and electrical properties of a MWNT/epoxy composite. Compos Sci Technol 62(15):1993–1998
52. Coleman JN, Khan U, Blau WJ, Gun'ko YK (2006) Small but strong: A review of the mechanical properties of carbon nanotube-polymer composites. Carbon 44(9):1624–1652
53. Coleman JN, Khan U, Gun'ko YK (2006) Mechanical reinforcement of polymers using carbon nanotubes. Adv Mater 18(6):689–706
54. Ma P-C, Siddiqui NA, Marom G, Kim J-K (2010) Dispersion and functionalization of carbon nanotubes for polymer-based nanocomposites: A review. Compos A Appl Sci Manuf 41(10):1345–1367
55. Miyagawa H, Misra M, Mohanty AK (2005) Mechanical properties of carbon nanotubes and their polymer nanocomposites. J Nanosci Nanotechnol 5(10):1593–1615
56. Moniruzzaman M, Winey KI (2006) Polymer nanocomposites containing carbon nanotubes. Macromolecules 39(16):5194–5205
57. Shim BS, Zhu J, Jan E, Critchley K, Ho S, Podsiadlo P, Sun K, Kotov NA (2009) Multiparameter structural optimization of single-walled carbon nanotube composites: toward record strength, stiffness, and toughness. ACS Nano 3(7):1711–1722
58. Spitalsky Z, Tasis D, Papagelis K, Galiotis C (2010) Carbon nanotube-polymer composites: Chemistry, processing, mechanical and electrical properties. Prog Polym Sci 35(3):357–401
59. Tsubokawa N (2005) Preparation and properties of polymer-grafted carbon nanotubes and nanofibers. Polymer 37(9):637–655
60. Dong YB, Ding J, Wang J, Fu X, Hu HM, Li S, Yang HB, Xu CS, Du ML, Fu YQ (2013) Synthesis and properties of the vapour-grown carbon nanofiber/epoxy shape memory and conductive foams prepared via latex technology. Compos Sci Technol 76:8–13
61. Palmeri MJ, Putz KW, Ramanathan T, Brinson LC (2011) Multi-scale reinforcement of CFRPs using carbon nanofibers. Compos Sci Technol 71(2):79–86
62. Adhikari AR, Partida E, Petty TW, Jones R, Lozano K, Guerrero C (2009) Fracture toughness of vapor grown carbon nanofiber-reinforced polyethylene composites. J Nanomater. doi:10.1155/2009/101870

63. Al-Saleh MH, Sundararaj U (2009) A review of vapor grown carbon nanofiber/polymer conductive composites. Carbon 47(1):2–22
64. Al-Saleh MH, Sundararaj U (2011) Review of the mechanical properties of carbon nanofiber/polymer composites. Compos A Appl Sci Manuf 42(12):2126–2142
65. Chirila V, Marginean G, Iclanzan T, Merino C, Brandl W (2007) Method for modifying mechanical properties of carbon nano-fiber polymeric composites. J Thermoplast Compos Mater 20(3):277–289
66. Choi YK, Sugimoto KI, Song SM, Gotoh Y, Ohkoshi Y, Endo M (2005) Mechanical and physical properties of epoxy composites reinforced by vapor grown carbon nanofibers. Carbon 43(10):2199–2208
67. Hammel E, Tang X, Trampert M, Schmitt T, Mauthner K, Eder A, Pötschke P (2004) Carbon nanofibers for composite applications. Carbon 42(5–6):1153–1158
68. Ma H, Zeng J, Realff ML, Kumar S, Schiraldi DA (2003) Processing, structure, and properties of fibers from polyester/carbon nanofiber composites. Compos Sci Technol 63(11):1617–1628
69. Rodriguez AJ, Guzman ME, Lim C-S, Minaie B (2011) Mechanical properties of carbon nanofiber/fiber-reinforced hierarchical polymer composites manufactured with multiscale-reinforcement fabrics. Carbon 49(3):937–948
70. Batra RC, Sears A (2007) Continuum models of multi-walled carbon nanotubes. Int J Solids Struct 44(22–23):7577–7596
71. Li G, Jones N (2007) Development of rubberized syntactic foam. Compos A Appl Sci Manuf 38(6):1483–1492
72. Jones N, Guoqiang L (2008) A CaO enhanced rubberized syntactic foam. Compos A Appl Sci Manuf 39(9):1404–1411
73. Maharsia R, Gupta N, Jerro HD (2006) Investigation of flexural strength properties of rubber and nanoclay reinforced hybrid syntactic foams. Mater Sci Eng A 417(1–2):249–258
74. Maharsia RR, Jerro HD (2007) Enhancing tensile strength and toughness in syntactic foams through nanoclay reinforcement. Mater Sci Eng A 454–455:416–422
75. Sunthonpagasit N, Duffey MR (2004) Scrap tires to crumb rubber: feasibility analysis for processing facilities. Resour Conserv Recycl 40(4):281–299
76. Shanmugharaj AM, Kim JK, Ryu SH (2005) UV surface modification of waste tire powder: Characterization and its influence on the properties of polypropylene/waste powder composites. Polym Test 24(6):739–745
77. Blumenthal MH (1994) Producing ground scrap tire rubber: A comparison between ambient and cryogenic technologies. Rubber Manufacturers Association, Washington, D.C, at: https://www.rma.org/getfile.cfm?ID=455&type=publication. Accessed on 4 August 2013
78. Wang J, Qin S (2007) Study on the thermal and mechanical properties of epoxy–nanoclay composites: The effect of ultrasonic stirring time. Mater Lett 61(19–20):4222–4224
79. Park JH, Jana SC (2003) Mechanism of exfoliation of nanoclay particles in epoxy–clay nanocomposites. Macromolecules 36(8):2758–2768
80. Hilding J, Grulke EA, Zhang ZG, Lockwood F (2003) Dispersion of carbon nanotubes in liquids. J Dispersion Sci Technol 24(1):1–41
81. Gupta N, Woldesenbet E (2004) Microballoon wall thickness effects on properties of syntactic foams. J Cell Plast 40(6):461–480
82. Kim HS, Plubrai P (2004) Manufacturing and failure mechanisms of syntactic foam under compression. Compos A Appl Sci Manuf 35(9):1009–1015
83. Gupta N, Ricci W (2006) Comparison of compressive properties of layered syntactic foams having gradient in microballoon volume fraction and wall thickness. Mater Sci Eng A 427(1–2):331–342
84. Gladysz GM, Perry B, McEachen G, Lula J (2006) Three-phase syntactic foams: structure-property relationships. J Mater Sci 41(13):4085–4092
85. Santhosh Kumar KS, Nair CPR, Ninan KN (2008) Mechanical properties of polybenzoxazine syntactic foams. J Appl Polym Sci 108(2):1021–1028

86. Shunmugasamy VC, Gupta N, Nguyen NQ, Coelho PG (2010) Strain rate dependence of damage evolution in syntactic foams. Mater Sci Eng A 527(23):6166–6177
87. Karthikeyan CS, Sankaran S, Kishore (2004) Elastic behaviour of plain and fibre-reinforced syntactic foams under compression. Mater Lett 58(6):995–999
88. Karthikeyan C, Murthy C, Sankaran S, Kishore (1999) Characterization of reinforced syntactic foams using ultrasonic imaging technique. Bull Mater Sci 22(4):811–815
89. Gupta N, Karthikeyan CS, Sankaran S, Kishore (1999) Correlation of processing methodology to the physical and mechanical properties of syntactic foams with and without fibers. Mater Charact 43(4):271–277
90. Wong JCH, Tervoort E, Busato S, Gauckler LJ, Ermanni P (2011) Controlling phase distributions in macroporous composite materials through particle-stabilized foams. Langmuir 27(7):3254–3260
91. Du Z, Bilbao-Montoya MP, Binks BP, Dickinson E, Ettelaie R, Murray BS (2003) Outstanding stability of particle-stabilized bubbles. Langmuir 19(8):3106–3108
92. Rodrigues JA, Rio E, Bobroff J, Langevin D, Drenckhan W (2011) Generation and manipulation of bubbles and foams stabilised by magnetic nanoparticles. Colloids Surf, A 384(1–3):408–416
93. Woldesenbet E, Sankella N (2009) Flexural properties of nanoclay syntactic foam sandwich structures. J Sandwich Struct Mater 11(5):425–444
94. Peter S, Woldesenbet E (2008) Nanoclay syntactic foam composites–High strain rate properties. Mater Sci Eng A 494(1–2):179–187
95. Gupta N, Nagorny R (2006) Tensile properties of glass microballoon-epoxy resin syntactic foams. J Appl Polym Sci 102(2):1254–1261
96. Wouterson EM, Boey FYC, Hu X, Wong S-C (2007) Effect of fiber reinforcement on the tensile, fracture and thermal properties of syntactic foam. Polymer 48(11):3183–3191
97. Huang Y-J, Vaikhanski L, Nutt SR (2006) 3D long fiber-reinforced syntactic foam based on hollow polymeric microspheres. Compos A Appl Sci Manuf 37(3):488–496
98. Agarwal BD, Broutman LJ, Chandrashekhara K (2006) Analysis and performance of fiber composites. Wiley, New York
99. Ajayan P, Suhr J, Koratkar N (2006) Utilizing interfaces in carbon nanotube reinforced polymer composites for structural damping. J Mater Sci 41(23):7824–7829
100. Zhang H, Tang L-C, Zhang Z, Friedrich K, Sprenger S (2008) Fracture behaviours of in situ silica nanoparticle-filled epoxy at different temperatures. Polymer 49(17):3816–3825
101. Li X-F, Lau K-T, Yin Y-S (2008) Mechanical properties of epoxy-based composites using coiled carbon nanotubes. Compos Sci Technol 68(14):2876–2881
102. Dean JM, Verghese NE, Pham HQ, Bates FS (2003) Nanostructure toughened epoxy resins. Macromolecules 36(25):9267–9270
103. Asif A, Rao VL, Ninan KN (2010) Nanoclay reinforced thermoplastic toughened epoxy hybrid syntactic foam: Surface morphology, mechanical and thermo mechanical properties. Mater Sci Eng A 527(23):6184–6192
104. Ayatollahi MR, Shadlou S, Shokrieh MM, Chitsazzadeh M (2011) Effect of multi-walled carbon nanotube aspect ratio on mechanical and electrical properties of epoxy-based nanocomposites. Polym Test 30(5):548–556
105. Gojny FH, Wichmann MHG, Fiedler B, Schulte K (2005) Influence of different carbon nanotubes on the mechanical properties of epoxy matrix composites - A comparative study. Compos Sci Technol 65(15–16):2300–2313
106. Paradise M, Goswami T (2007) Carbon nanotubes - production and industrial applications. Mater Des 28(5):1477–1489
107. Esawi AMK, Farag MM (2007) Carbon nanotube reinforced composites: potential and current challenges. Mater Des 28(9):2394–2401
108. Sanchez M, Rams J, Campo M, Jimenez-Suarez A, Urena A (2011) Characterization of carbon nanofiber/epoxy nanocomposites by the nanoindentation technique. Compos B Eng 42(4):638–644

109. Yang S, Taha-Tijerina J, Serrato-Diaz Vn, Hernandez K, Lozano K (2007) Dynamic mechanical and thermal analysis of aligned vapor grown carbon nanofiber reinforced polyethylene. Compos B Eng 38(2):228–235

110. Zeng J, Saltysiak B, Johnson WS, Schiraldi DA, Kumar S (2004) Processing and properties of poly(methyl methacrylate)/carbon nanofiber composites. Compos B Eng 35(3):245–249

111. Dimchev M, Caeti R, Gupta N (2010) Effect of carbon nanofibers on tensile and compressive characteristics of hollow particle filled composites. Mater Des 31(3):1332–1337

112. Colloca M, Gupta N, Porfiri M (2013) Tensile properties of carbon nanofiber reinforced multiscale syntactic foams Composite Part B. Engineering 44(1):584–591

113. Endo M, Kim YA, Hayashi T, Yanagisawa T, Muramatsu H, Ezaka M, Terrones H, Terrones M, Dresselhaus MS (2003) Microstructural changes induced in stacked cup carbon nanofibers by heat treatment. Carbon 41(10):1941–1947

114. Palmeri MJ, Putz KW, Brinson LC (2010) Sacrificial bonds in stacked-cup carbon nanofibers: Biomimetic toughening mechanisms for composite systems. ACS Nano 4(7):4256–4264

115. Wouterson EM, Boey FYC, Wong SC, Chen L, Hu X (2007) Nano-toughening versus micro-toughening of polymer syntactic foams. Compos Sci Technol 67(14):2924–2933

116. Wang K, Chen L, Wu J, Toh ML, He C, Yee AF (2005) Epoxy nanocomposites with highly exfoliated clay: mechanical properties and fracture mechanisms. Macromolecules 38(3):788–800

117. Gojny FH, Nastalczyk J, Roslaniec Z, Schulte K (2003) Surface modified multi-walled carbon nanotubes in CNT/epoxy-composites. Chem Phys Lett 370(5–6):820–824

118. Santhosh Kumar KS, Nair CPR, Ninan KN (2008) Silica fiber-polybenzoxazine-syntactic foams; processing and properties. J Appl Polym Sci 107(2):1091–1099

119. Vaikhanksi L, Nutt SR (2003) Synthesis of composite foam from thermoplastic microspheres and 3D long fibers. Compos A Appl Sci Manuf 34(8):755–763

120. Wouterson EM, Boey FYC, Hu X, Wong SC (2007) Effect of fiber reinforcement on the tensile, fracture and thermal properties of syntactic foam. Polymer 48(11):3183–3191

121. John B, Nair CPR, Ninan KN (2010) Effect of nanoclay on the mechanical, dynamic mechanical and thermal properties of cyanate ester syntactic foams. Mater Sci Eng A 527(21–22):5435–5443

122. Rizzi E, Papa E, Corigliano A (2000) Mechanical behavior of a syntactic foam: experiments and modeling. Int J Solids Struct 37(40):5773–5794

123. Xu W, Li G (2010) Constitutive modeling of shape memory polymer based self-healing syntactic foam. Int J Solids Struct 47(9):1306–1316

124. Marur PR (2005) Effective elastic moduli of syntactic foams. Mater Lett 59(14–15):1954–1957

125. Bardella L, Genna F (2001) On the elastic behavior of syntactic foams. Int J Solids Struct 38(40–41):7235–7260

126. Nguyen NQ, Gupta N (2010) Analyzing the effect of fiber reinforcement on properties of syntactic foams. Mater Sci Eng A 527(23):6422–6428

127. Aureli M, Porfiri M, Gupta N (2010) Effect of polydispersivity and porosity on the elastic properties of hollow particle filled composites. Mech Mater 42(7):726–739

128. Tagliavia G, Porfiri M, Gupta N (2011) Elastic interaction of interfacial spherical-cap cracks in hollow particle filled composites. Int J Solids Struct 48(7–8):1141–1153

129. Tagliavia G, Porfiri M, Gupta N (2011) Analysis of particle-to-particle elastic interactions in syntactic foams. Mech Mater 43(12):952–968

130. Christensen RM (1979) Mechanics of composite materials. Dover Publications, Mineola

131. Torquato S (2001) Random heterogeneous materials: microstructure and macroscopic properties. Springer, New York

132. Pal R (2005) New models for effective Young's modulus of particulate composites. Compos B Eng 36(6–7):513–523

133. Bardella L, Sfreddo A, Ventura C, Porfiri M, Gupta N (2012) A critical evaluation of micromechanical models for syntactic foams on the basis of three-dimensional finite element analyses. Mech Mater 50:53–69

134. Barhdadi EH, Lipinski P, Cherkaoui M (2007) Four phase model: A new formulation to predict the effective elastic moduli of composites. J Eng Mater Technol 129(2):313–320

135. Lutz MP, Zimmerman RW (2005) Effect of an inhomogeneous interphase zone on the bulk modulus and conductivity of a particulate composite. Int J Solids Struct 42(2):429–437
136. Pal R (2005) Modeling viscoelastic behavior of particulate composites with high volume fraction of filler. Mater Sci Eng A 412(1–2):71–77
137. Tagliavia G, Porfiri M, Gupta N (2010) Analysis of flexural properties of hollow-particle filled composites. Compos B Eng 41(1):86–93
138. Vaccarini L, Desarmot G, Almairac R, Tahir S, Goze C, Bernier P (2000) Reinforcement of an epoxy resin by single walled nanotubes. AIP Conf Proc 544(1):521–525
139. Zeng QH, Yu AB, Lu GQ (2008) Multiscale modeling and simulation of polymer nanocomposites. Prog Polym Sci 33(2):191–269
140. Griebel M, Hamaekers J (2004) Molecular dynamics simulations of the elastic moduli of polymer - carbon nanotube composites. Comput Methods Appl Mech Eng 193(17–20):1773–1788
141. Gupta SS, Bosco FG, Batra RC (2010) Wall thickness and elastic moduli of single-walled carbon nanotubes from frequencies of axial, torsional and inextensional modes of vibration. Comput Mater Sci 47(4):1049–1059
142. Mylvaganam K, Zhang LC (2004) Important issues in a molecular dynamics simulation for characterising the mechanical properties of carbon nanotubes. Carbon 42(10):2025–2032
143. Frankland SJV, Caglar A, Brenner DW, Griebel M (2002) Molecular simulation of the influence of chemical cross-links on the shear strength of carbon nanotube−polymer interfaces. J Phys Chem B 106(12):3046–3048
144. Frankland SJV, Harik VM, Odegard GM, Brenner DW, Gates TS (2003) The stress–strain behavior of polymer–nanotube composites from molecular dynamics simulation. Compos Sci Technol 63(11):1655–1661
145. Bunn P, Mottram JT (1993) Manufacture and compression properties of syntactic foams. Composites 24(7):565–571
146. Gupta N, Kishore, Woldesenbet E, Sankaran S (2001) Studies on compressive failure features in syntactic foam material. J Mater Sci 36(18):4485–4491
147. Gupta N, Maharsia R, Dwayne Jerro H (2005) Enhancement of energy absorption characteristics of hollow glass particle filled composites by rubber addition. Mater Sci Eng A 395(1–2):233–240
148. Li G, John M (2008) A crumb rubber modified syntactic foam. Mater Sci Eng A 474(1–2):390–399
149. Gupta N, Maharsia R (2005) Enhancement of energy absorption in syntactic foams by nanoclay incorporation for sandwich core applications. Appl Compos Mater 12(3):247–261
150. Guzman ME, Rodriguez AJ, Minaie B, Violette M (2012) Processing and properties of syntactic foams reinforced with carbon nanotubes. J Appl Polym Sci 124(3):2383–2394
151. Poveda R, Dorogokupets G, Gupta N (2013) Quasi-static and high strain rate compressive response of carbon nanofiber reinforced composites. unpublished results
152. Naik NK, Shankar PJ, Kavala VR, Ravikumar G, Pothnis JR, Arya H (2011) High strain rate mechanical behavior of epoxy under compressive loading: Experimental and modeling studies. Mater Sci Eng A 528(3):846–854
153. Woldesenbet E, Peter S (2009) Volume fraction effect on high strain rate properties of syntactic foam composites. J Mater Sci 44(6):1528–1539
154. Li P, Petrinic N, Siviour CR, Froud R, Reed JM (2009) Strain rate dependent compressive properties of glass microballoon epoxy syntactic foams. Mater Sci Eng A 515(1–2):19–25
155. Karthikeyan CS, Sankaran S (2001) Effect of absorption in aqueous and hygrothermal media on the compressive properties of glass fiber reinforced syntactic foam. J Reinf Plast Compos 20(11):982–993
156. Gu H (2009) Dynamic mechanical analysis of the seawater treated glass/polyester composites. Mater Des 30(7):2774–2777
157. Karthikeyan CS, Sankaran S, Jagdish Kumar MN, Kishore (2001) Processing and compressive strengths of syntactic foams with and without fibrous reinforcements. J Appl Polym Sci 81(2):405–411

158. Karthikeyan CS, Sankaran S (2000) Comparison of compressive properties of fiber-free and fiber-bearing syntactic foams. J Reinf Plast Compos 19(9):732–742

159. Karthikeyan CS, Sankaran S, Kishore (2005) Flexural behaviour of fibre-reinforced syntactic foams. Macromol Mater Eng 290(1):60–65

160. Karthikeyan CS, Sankaran S, Kishore (2000) Influence of chopped strand fibres on the flexural behaviour of a syntactic foam core system. Polym Int 49(2):158–162

161. Karthikeyan CS, Sankaran S, Kishore (2007) Investigation of bending modulus of fiber-reinforced syntactic foams for sandwich and structural applications. Polym Adv Technol 18(3):254–256

162. Yusriah L, Mariatti M (2013) Effect of hybrid phenolic hollow microsphere and silica-filled vinyl ester composites. J Compos Mater 47(2):169–182

163. Saha MC, Nilufar S (2010) Nanoclay-reinforced syntactic foams: Flexure and thermal behavior. Polym Compos 31(8):1332–1342

164. Ferreira JAM, Capela C, Costa JD (2010) A study of the mechanical behaviour on fibre reinforced hollow microspheres hybrid composites. Compos A Appl Sci Manuf 41(3):345–352

165. Yusriah L, Mariatti M, Abu Bakar A (2010). The properties of vinyl ester composites reinforced with different types of woven fabric and hollow phenolic microspheres. J Reinf Plast Compos 29(20):3066–3073

166. Zhang L, Ma J (2010) Effect of coupling agent on mechanical properties of hollow carbon microsphere/phenolic resin syntactic foam. Compos Sci Technol 70(8):1265–1271

167. Wouterson EM, Boey FYC, Hu X, Wong S-C (2005) Specific properties and fracture toughness of syntactic foam: Effect of foam microstructures. Compos Sci Technol 65(11–12):1840–1850

168. Zhang L, Ma J (2013) Effect of carbon nanofiber reinforcement on mechanical properties of syntactic foam. Mater Sci Eng A 574:191–196

169. Menard KP (1999) Dynamic mechanical analysis a practical introduction. CRC Press, Boca Raton

170. Ferreira JAM, Capela C, Costa JD (2011) Dynamic mechanical analysis of hybrid fiber/glass microspheres composites. Strain 47(3):275–280

171. Shunmugasamy VC, Pinisetty D, Gupta N (2013) Viscoelastic properties of hollow glass particle filled vinyl ester matrix syntactic foams: effect of temperature and loading frequency. J Mater Sci 48(4):1685–1701

172. Capela C, Ferreira JAM, Costa JD (2010) Viscoleastic properties assessment of syntactic foams by dynamic mechancial analysis. Mater Sci Forum 636–637:280–286

173. Peter SL, Woldesenbet E (2009) Nanoclay and microballoons wall thickness effect on dynamic properties of syntactic foam. J Eng Mater Technol 131(2):021007

174. Wehmer P. (2008) High strain rate characteristics of rubber modified syntactic foams, in Department of Mechanical Engineering. MS Thesis, Louisiana State University, Baton Rouge. p 57 http://etd.lsu.edu/docs/available/etd-11132008-130753/